I0467369

An In-Depth Analysis of Stochastic Kronecker Graphs

C. SESHADHRI, ALI PINAR, and TAMARA G. KOLDA, Sandia National Laboratories

Graph analysis is playing an increasingly important role in science and industry. Due to numerous limitations in sharing real-world graphs, models for generating massive graphs are critical for developing better algorithms. In this article, we analyze the stochastic Kronecker graph model (SKG), which is the foundation of the Graph500 supercomputer benchmark due to its favorable properties and easy parallelization. Our goal is to provide a deeper understanding of the parameters and properties of this model so that its functionality as a benchmark is increased. We develop a rigorous mathematical analysis that shows this model cannot generate a power-law distribution or even a lognormal distribution. However, we formalize an enhanced version of the SKG model that uses random noise for smoothing. We prove both in theory and in practice that this enhancement leads to a lognormal distribution. Additionally, we provide a precise analysis of isolated vertices, showing that the graphs that are produced by SKG might be quite different than intended. For example, between 50% and 75% of the vertices in the Graph500 benchmarks will be isolated. Finally, we show that this model tends to produce extremely small core numbers (compared to most social networks and other real graphs) for common parameter choices.

Categories and Subject Descriptors: D.2.8 [**Software Engineering**]: Metrics—*Complexity measures, performance measures*; E.1 [**Data**]: Data Structures—*Graphs and networks*

General Terms: Algorithms, Theory

Additional Key Words and Phrases: graph models, R-MAT, Stochastic Kronecker Graphs (SKG), Graph500

ACM Reference Format:
Seshadhri, C., Pinar, A., and Kolda, T. G. 2013. An in-depth analysis of stochastic Kronecker graphs. J. ACM 60, 2, Article 13 (April 2013), 32 pages.
DOI: http://dx.doi.org/10.1145/2450142.2450149

1. INTRODUCTION

The role of graph analysis is becoming increasingly important in science and industry because of the prevalence of graphs in diverse scenarios such as social networks, the Web, power grid networks, and even scientific collaboration studies. Massive graphs occur in a variety of situations, and we need to design better and faster algorithms in order to study them. However, it can be difficult to access to informative large graphs in order to test our algorithms. Companies like Netflix, AOL, and Facebook have vast

This work was funded by the applied mathematics program at the United States Department of Energy and performed at Sandia National Laboratories, a multiprogram laboratory operated by Sandia Corporation, a wholly owned subsidiary of Lockheed Martin Corporation, for the United States Department of Energy's National Nuclear Security Administration under contract DE-AC04-94AL85000.
Authors' address: Sandia National Laboratories, Livermore, CA 94551, Correspondence email: apinar@sandia.gov.

arrays of data but cannot share it due to legal or copyright issues.[1] Moreover, graphs with billions of vertices cannot be communicated easily due to their sheer size.

As was noted in Chakrabarti and Faloutsos [2006], good *graph models* are extremely important for the study and algorithmics of real networks. Such a model should be fairly easy to implement and have few parameters, while exhibiting the common properties of real networks. Furthermore, models are needed to test algorithms and architectures designed for large graphs. But the theoretical and research benefits are also obvious: gaining insight into the properties and processes that create real networks.

The *stochastic Kronecker graph* (SKG) [Leskovec and Faloutsos 2007; Leskovec et al. 2010], a generalization of the *recursive matrix* (R-MAT) model [Chakrabarti et al. 2004], has been proposed for these purposes. It has very few parameters and can generate large graphs quickly. Indeed, it is one of the few models that can generate graphs fully in *parallel*. It has been empirically observed to have interesting real-network-like properties. We stress that this is not just of theoretical or academic interest—this model has been chosen to create graphs for the Graph500 supercomputer benchmark [Graph500 Steering Committee 2010].

It is important to know how the parameters of this model affect various properties of the graphs. We stress that a mathematical analysis is important for understanding the inner working of a model. We quote Mitzenmacher [2006]: "I would argue, however, that without validating a model it is not clear that one understands the underlying behavior and therefore how the behavior might change over time. It is not enough to plot data and demonstrate a power law, allowing one to say things about current behavior; one wants to ensure that one can accurately predict future behavior appropriately, and that requires understanding the correct underlying model."

1.1. Notation and Background

We explain the SKG model and notation. Our goal is to generate a directed graph $G = (V, E)$ with $n = |V|$ nodes and $m = |E|$ edges. The general form of the SKG model allows for an arbitrary square generator matrix and assumes that n is a power of its size. Here, we focus on the 2×2 case (which is equivalent to R-MAT), defining the generating matrix as

$$T = \begin{bmatrix} t_1 & t_2 \\ t_3 & t_4 \end{bmatrix} \quad \text{with} \quad t_1 + t_2 + t_3 + t_4 = 1 \text{ and } \min_i t_i > 0.$$

We assume that $n = 2^\ell$ for some integer $\ell > 0$. For the sake of cleaner formulae, we assume that ℓ is even in our analyses. Each edge is inserted according to the probabilities defined by

$$P = \underbrace{T \otimes T \otimes \cdots \otimes T}_{\ell \text{ times}},$$

where $\otimes$ denotes the Kronecker product operation. In practice, the matrix P is never formed explicitly. Instead, each edge is inserted as follows. Divide the adjacency matrix into four quadrants, and choose one of them with the corresponding probability t_1, t_2, t_3, or t_4. Once a quadrant is chosen, repeat this recursively in that quadrant. Each time we iterate, we end up in a square submatrix whose dimensions are exactly halved. After ℓ iterations, we reach a single cell of the adjacency matrix, and an edge is inserted. It should be noted that here we take a slight liberty in requiring the entries of T to sum to 1. In fact, the SKG model as defined in Leskovec et al. [2010] works with the matrix

---

[1]For example, Netflix opted not to pursue the Netflix Prize sequel due to concerns about lawsuits; see http://blog.netflix.com/2010/03/this-is-neil-hunt-chief-product-officer.html.

Table I. Parameters for SKG Models

PRIMARY PARAMETERS
$-T = \begin{bmatrix} t_1 & t_2 \\ t_3 & t_4 \end{bmatrix}$ = generating matrix with $t_1 + t_2 + t_3 + t_4 = 1$
$-\ell$ = number of levels (assumed even for analysis)
$-m$ = number of edges
DERIVATIVE PARAMETERS
$-n = 2^\ell$ = number of nodes
$-\Delta = m/n$ = average degree
$-\sigma = t_1 + t_2 - 0.5$ = skew

mP, which is considered the matrix of probabilities for the existence of each individual edge (though it might be more accurate to think of it as an expected value).

Note that all edges can be inserted in parallel. This is one of the major advantages of the SKG model and why it is appropriate for generating large supercomputer benchmarks.

For convenience, we also define some derivative parameters that will be useful in subsequent discussions. We let $\Delta = m/n$ denote the *average degree* and let $\sigma = t_1 + t_2 - 0.5$ denote the *skew*. The parameters of the SKG model are summarized in Table 1.

1.2. Our Contributions

Our overall contribution is to provide a thorough study of the properties of SKG and show how the parameters affect these properties. We focus on the degree distribution, the number of (non-isolated nodes), the core sizes, and the trade-offs in these various goals. We give rigorous mathematical theorems and proofs explaining the degree distribution of SKG, a noisy version of SKG, and the number of isolated vertices.

(1) *Degree Distribution*. We provide a rigorous mathematical analysis of the degree distribution of SKG. The degree distribution has often been claimed to be power-law, or sometimes lognormal [Chakrabarti et al. 2004; Leskovec et al. 2010; Kim and Leskovec 2010]. Kim and Leskovec [2010] prove that the degree distribution has some lognormal characteristics. Groër et al. [2011] give exact series expansions for the degree distribution, and express it as a mixture of normal distributions. This provides a qualitative explanation for the oscillatory behavior of the degree distribution (refer to Figure 1). Since the distribution is quite far from being truly lognormal, there has been no simple closed form expression that closely approximates it. We fill this gap by providing a complete mathematical description. We prove that SKG cannot generate a power law distribution, or even a lognormal distribution. It is most accurately characterized as fluctuating between a lognormal distribution and an exponential tail. We provide a simple formula that approximates the degree distribution.

(2) *Noisy SKG*. It has been mentioned in passing [Chakrabarti et al. 2004] that adding noise to SKG at each level smoothens the degree distribution, but this has never been formalized or studied. We define a specific noisy version of SKG (NSKG). We prove theoretically and empirically that NSKG leads to a lognormal distribution. (We give some experimental results showing a naive addition of noise does not work.) The lognormal distribution is important since it has been observed in real data [Bi et al. 2001; Pennock et al. 2002; Mitzenmacher 2003; Clauset et al. 2009]. One of the major benefits of our enhancement is that only ℓ additional random numbers are needed in total. Using Graph500 parameters, Figure 1 plots the degree distribution of a (standard) SKG and NSKG for two levels of (maximum) noise. We can clearly see that noise dampens the oscillations, leading to a lognormal distribution.

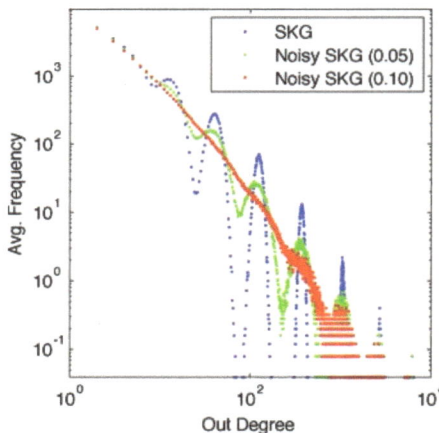

Fig. 1. Comparison of degree distributions (averaged over 25 instances) for SKG and two noisy variations, using the T from the Graph500 Benchmark parameters with $\ell = 16$.

Table II.
Expected percentage of isolated vertices and repeat edges, along with average degree of *nonisolated* nodes for the Graph500 benchmark. Excluding the isolated vertices results in a much higher average degree than the value of 16 that is specified by the benchmark.

ℓ	% Isolated Nodes	% Repeat Edges	Avg. Degree
26	51	1.2	32
29	57	0.7	37
32	62	0.4	41
36	67	0.2	49
39	71	0.1	55
42	74	0.1	62

We note that though the modification of NSKG is straightforward, the reason why it works is not. It involves an intricate mathematical analysis, which may be of theoretical interest in itself.

(3) *Isolated Vertices.* An isolated vertex is one that has no edges incident to it (and hence is not really part of the output graph). We provide a formula that accurately estimates the fraction of isolated vertices. We discover the surprising result that in the Graph500 benchmark graphs, 50–75% vertices are isolated; see Table II. This is a major concern for the benchmark, since the massive graph generated has a much reduced size. Furthermore, the average degree is now much higher than expected.

(4) *Core Numbers.* The study of k-cores is an important tool used to study the structure of social networks because it is a mark of the connectivity and special processes that generate these graphs [Chakrabarti and Faloutsos 2006; Kumar et al. 2010; Alvarez-Hamelin et al. 2008; Gkantsidis et al. 2003; Goltsev et al. 2006; Carmi et al. 2007; Andersen and Chellapilla 2009]. We empirically show how the core numbers have unexpected correlations with SKG parameters. We observed that for most of the current SKG parameters used for modeling real graphs, max core numbers are extremely small (much smaller than most corresponding real graphs). We show how modifying the matrix T affects core numbers. Most strikingly, we observe that changing T to increase the max core number actually leads to an increase in the fraction of isolated vertices.

1.3. Influence on Graph500 Benchmark

Our results have been communicated to the Graph500 steering committee, who have found them useful in understanding the Graph500 benchmark. The oscillations in the degree distribution of SKG was a major concern for the committee. Our proposed NSKG model has been implemented in the current Graph500 code.[2]

Our analysis also solves the mystery of isolated vertices and how they are related to the SKG parameters. Members of the steering committee had observed that the number of isolated vertices varied greatly with the matrix T, but did not have an explanation for this.

1.4. Parameters for Empirical Study

Throughout the article, we discuss a few sets of SKG parameters. The first is the Graph500 benchmark [Graph500 Steering Committee 2010]. The other two are parameters used in [Leskovec et al. 2010] to model a co-authorship network (CAHepPh) and a web graph (WEB NotreDame). We list these parameters here for later reference.

—*Graph500*: $T = [0.57, 0.19; 0.19, 0.05]$, $\ell \in \{26, 29, 32, 36, 39, 42\}$, and $m = 16 \cdot 2^\ell$.
—*CAHepPh*: $T = [0.42, 0.19; 0.19, 0.20]$, $\ell = 14$, and $m = 237, 010$.
—*WEBNotreDame*[3]: $T = [0.48, 0.20; 0.21, 0.11]$, $\ell = 18$, and $m = 1, 497, 134$.

2. PREVIOUS WORK

The R-MAT model was defined by Chakrabarti et al. [2004]. The general and more powerful SKG model was introduced by Leskovec et al. [2005] and fitting algorithms were proposed by Leskovec and Faloutsos [2007] (combined in Leskovec et al. [2010]). This model has generated significant interest and notably was chosen for the Graph500 benchmark [Graph500 Steering Committee 2010]. Kim and Leskovec [2010] defined the Multiplicative Attribute Graph (MAG) model, a generalization of SKG where each level may have a different matrix T. They suggest that certain configurations of these matrices could lead to power-law distributions.

Since the appearance of the SKG model, there have been analyses of its properties. The original paper [Leskovec et al. 2010] provides some basic theorems and empirically show a variety of properties. Mahdian and Xu [2011] specifically study how the model *parameters* affect the graph properties. They show phase transition behavior (asymptotically) for occurrence of a large connected component and shrinking diameter. They also initiate a study of isolated vertices. When the SKG parameters satisfy a certain condition, the number of isolated vertices approaches n; however, their theorems do not help predict the number of isolated vertices for a given setting of SKG. In the analysis of the MAG model [Kim and Leskovec 2010], it is shown that the SKG degree distribution has some lognormal characteristics. (Lognormal distributions have been observed in real data [Bi et al. 2001; Pennock et al. 2002; Clauset et al. 2009]. Mitzenmacher [2003] gives a survey of lognormal distributions.)

Sala et al. [2010] perform an extensive empirical study of properties of graph models, including SKG. Miller et al. [2010] show that they can detect anomalies embedded in an SKG. Moreno et al. [2010] study the distributional properties of families of SKG.

As noted in Chakrabarti et al. [2004], the SKG generation procedure may give repeated edges. Hence, the number of edges in the graph differs slightly from the number

---

[2]The file generator/graph-generator.c in the most recent version as of July 2012 (2.1.4) has the implementation, with a variable SPK_NOISE_LEVEL controlling the NSKG noise. Available at http//www.graph500.org/sites/default/files/files/graph500-2.1.4.tar.bz2

[3]In Leskovec et al. [2010], ℓ was 19. We make it even because, for the sake of presentation, we perform experiments and derive formulas for even ℓ.

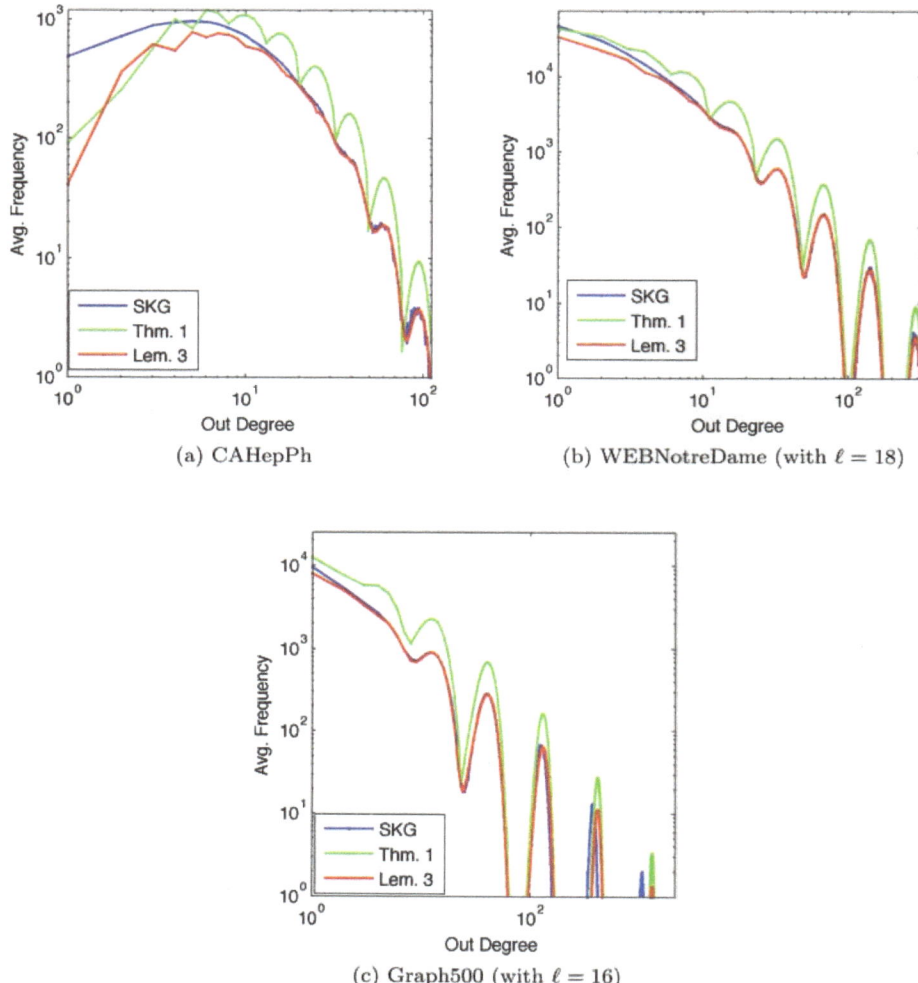

Fig. 2. We plot the degree distribution of graphs generated using our three different SKG parameter sets. We then plot the respective bounds predicted by Theorem 3.1 and Lemma 3.2. Observe how Theorem 3.1 correctly guesses the peaks and troughs of the degree distribution. Lemma 3.2 is practically an exact match (except when the degree is below 2ℓ or, in Graph500, slight inaccuracies when the degree is too large).

of insertions (though, in practice, this is barely 1% for Graph500). Groër et al. [2011] prove that the number of vertices of a given degree is asymptotically normally distributed, and provide algorithms to compute the expected number of edges in the graph (as a function of the number of insertions) and the expected degree distribution.

3. DEGREE DISTRIBUTION

In this section, we analyze the degree distribution of SKG, which is known to follow a multinomial distribution. While an exact expression for this distribution can be written, this is unfortunately a complicated sum of binomial coefficients. Studying the log-log plots of the degree distribution, one sees a general heavy-tail like behavior, but there are large oscillations. The degree distribution is not monotonically decreasing.

Refer to Figure 2 for some examples of SKG degree distributions (plotted in log-log scale). Groër et al. [2011] show that the degree distribution behaves like the sum of Gaussians, giving some intuition for the oscillations. Recent work of Kim and Leskovec [2010] provide some mathematical analysis explaining connections to a lognormal distribution. But many questions remain. What does the distribution oscillate between? Is the distribution bounded below by a power law? Can we approximate the distribution with a simple closed form function? None of these questions have satisfactory answers.

Our analysis gives a precise explanation for the SKG degree distribution. We prove that the SKG degree distribution oscillates between a lognormal and exponential tail. We provide plots and experimental results to support more intuition for our theorems.

The oscillations are a disappointing feature of SKG. Real degree distributions do not have large oscillations (to the contrary, they are monotonically decreasing), and more importantly, do not have any exponential tail behavior. This is a major issue both for modeling and benchmarking purposes since degree distribution is one of the primary characteristics that distinguishes real networks.

In order to rectify the oscillations, we apply a certain model of noise and provide both mathematical and empirical evidence that this "straightens out" the degree distribution. This is discussed in Section 4. Indeed, small amounts of noise lead to a degree distribution that is predominantly lognormal. This also shows an appealing aspect of our degree distribution analysis. We can naturally explain how noise affects the degree distribution and give explicit bounds on these affects.

We make a caveat here. Technically, the SKG model creates *multigraphs*, since there can be repeated edges. Our theorems and expressions will deal with degree distributions of this multigraph. Conventionally, this is reduced to a simple graph by removing repeated edges. Groër et al. [2011] give details expressions and explanations relating the degree distributions on the multigraph and the induced simple graph. Our empirical results show that for a variety of parameters (including the Graph 500 setting), our theorems match the degree distribution of the underlying simple graph. Simple graphs are used in all empirical studies.

3.1. Notation

The ℓ-bit binary representation of the vertices, numbered 0 to $n-1$, provides a straightforward way to partition the vertices. Specifically, each vertex has a binary representation and therefore corresponds to an element of the boolean hypercube $\{0, 1\}^{\ell}$. We can partition the vertices into *slices*, where each slice consists of vertices whose representations have the same number of zeros.[4] Recall that we assume ℓ is even. For $r \in [-\ell/2, \ell/2]$, we say that *slice* r, denoted S_r, consists of all vertices whose binary representations have exactly $(\ell/2 + r)$ zeros.

These binary representations and slices are intimately connected with edge insertions in the SKG model. For each insertion, we are trying to randomly choose a source-sink pair. First, let us simply choose the first bit (of the representations) of the source and the sink. Note that there are 4 possibilities (first bit for source, second for sink): 00, 01, 10, and 11. We choose one of the combinations with probabilities t_1, t_2, t_3, and t_4 respectively. This fixes the first bit of the source and sink. We perform this procedure again to choose the second bit of the source and sink. Repeating ℓ times, we finally decide the source and sink of the edge. Note that as $|r|$ becomes smaller, a vertex in an r-slice tends to have a higher degree.

---

[4] There are usually referred to as the *levels of the Boolean hypercube*. In the SKG literature, *levels* is used to refer to ℓ, and hence we use a different term.

Table III. Parameters for Analysis of SKG Models

GENERAL QUANTITIES

—$\tau = (1 + 2\sigma)/(1 - 2\sigma)$
—$\lambda = \Delta(1 - 4\sigma^2)^{\ell/2}$
—$r \in \{-\ell/2, \ldots, \ell/2\}$ denotes a slice index
—d denotes a degree (typically assumed $< \sqrt{n}$)
—$\deg(v) =$ outdegree of node v
—$S_r =$ set of nodes whose binary representation have exactly $\ell/2 + r$ zeros

QUANTITIES ASSOCIATED WITH DEGREE d

—$X_d =$ random variable for the number of vertices of outdegree d
—$\theta_d = \ln(d/\lambda)/\ln\tau$
—$\Gamma_d = \lfloor\theta_d\rceil$ (nearest integer to θ_d)
—$\gamma_d = |\theta_d - \Gamma_d| \in [0, 0.5]$
—$r_d = \lfloor\theta_d\rfloor$ (only interesting for $r_d < \ell/2$)
—$\delta_d = \theta_d - r_d$

For a real number x, we use $\lfloor x \rceil$ to denote the closest integer to x. There are certain quantities that will be important in our analysis. These are summarized in Table III.

Our results are fundamentally asymptotic in nature, so we explain the assumptions on T and the implicit assumptions of our results. We assume T to be a *fixed* matrix with the following conditions. All entries are positive and strictly less than 1. The number t_1 is the largest entry, and $\min(t_1 + t_2, t_1 + t_3) > 1/2$. This ensures that $\sigma \in (0, 1/2)$, τ is positive and finite, and λ is non-zero. We want to note that these conditions are satisfied by all SKG parameters that have been used to generate realistic graph instances, to the best of our knowledge. Indeed, when $\sigma = 1/2$, the degree distribution is Poisson.

We fix the matrix T and average degree $\Delta > 1$, and think of ℓ as increasing. The asymptotics hold for an increasing ℓ. Note that since $n = 2^\ell$, this means that n and m are also increasing. We use $o(1)$ as a shorthand for a quantity that is negligible as $\ell \to \infty$. Typically, this converges to zero rapidly as ℓ increases. Given two quantities or expressions A and B, $A = (1 \pm o(1))B$ will be shorthand for $A \in [(1 - o(1))B, (1 + o(1))B]$.

As we mentioned earlier, all our results are for the SKG multigraph. For convenience, we will just refer to this a graph.

3.2. Explicit Formula for Degree Distribution

We begin by stating and explaining the main result of this section. To provide clean expressions, we make certain approximations which are slightly off for certain regions of d and ℓ (essentially, when d is either too small or too large). Our main technical result is Lemma 3.2, which gives a tight expression for the degree distribution. A more interpretable version is given first as Theorem 3.1, which is stated as an upper bound. The remainder of the section gives a proof for this, which can be skipped if the reader is only interested in the results. This theorem expresses the oscillations between the lognormal and exponential tail. The lower order error terms in all the following are extremely small.

We focus on outdegrees, but these theorems hold for indegrees as well. To make dependences clear, we remind the reader that the "free" variables are T, Δ, ℓ. The first two are fixed to constants, and ℓ is increasing. Hence, the asymptotics are over ℓ. All other parameters are functions of these quantities.

We begin by giving a more digestible form of our main result, stated in Theorem 3.1. The more precise version is given in Lemma 3.2. A reader interested in the general message can skip Lemma 3.2.

THEOREM 3.1. *Assume $d \in [(e \ln 2)\ell, \sqrt{n}]$. If $\Gamma_d \geq \ell/2$, then $\mathbf{E}[X_d]$ is negligible, that is, $o(1)$; otherwise, if $\Gamma_d < \ell/2$, then (up to an additive exponential tail)*

$$\mathbf{E}[X_d] \leq \frac{1}{\sqrt{d}} \exp\left(\frac{-d\gamma_d^2 \ln^2 \tau}{2}\right) \binom{\ell}{\ell/2 + \Gamma_d}.$$

Remark. This means that the expected outdegree distribution of a SKG is bounded above by a function that oscillates between a lognormal and an exponential tail.

Note that $\Gamma_d = \lfloor \ln(d/\lambda)/\ln \tau \rceil = \Theta(\ln d)$. Hence $\binom{\ell}{\ell/2+\Gamma_d}$ can be thought of as $\binom{\ell}{\ell/2+\Theta(\ln d)}$. The function $\binom{\ell}{\ell/2+x}$ represents an asymptotically normal distribution of x, and therefore $\binom{\ell}{\ell/2+\Gamma_d}$ is a *lognormal distribution of d*. This lognormal term is multiplied by $\exp(-d\gamma_d^2 \ln^2 \tau/2)$. By definition, $\gamma_d \in [0, 1/2]$. When γ_d is close to 0, then the exponential term is almost 1. Hence the product represents a lognormal tail. On the other hand, when γ_d is a constant (say > 0.2), then the product becomes an exponential tail. Observe that γ_d oscillates between 0 and 1/2, leading to the characteristic behavior of SKG. As θ_d becomes closer to an integer, there are more vertices of degree d. As it starts to have a larger fractional part, the number of vertices of degree d is bounded above by an exponential tail. Note that there are many values of d (a constant fraction) where $\gamma_d > 0.2$. Hence, for all these d, the degrees are bounded above by an exponential tail. As a result, the degree distribution cannot be a power law or a lognormal.

The estimates provided by Theorem 3.1 for our three different SKG parameter sets are shown in Figure 2. Note how this simple estimate matches the oscillations of the actual degree distribution accurately.

We provide a more complex expression in Lemma 3.2 that almost completely explains the degree distribution. Theorem 3.1 is a direct corollary of this lemma. In the following, the expectation is over the random choice of the graph.

LEMMA 3.2. *For SKG, assume $d \in [(e \ln 2)\ell, \sqrt{n}]$. If $r_d \geq \ell/2$, $\mathbf{E}[X_d]$ is negligible; otherwise, we have*

$$\mathbf{E}[X_d] = \frac{1 \pm o(1)}{\sqrt{2\pi d}} \left\{ \exp\left(\frac{-d\delta_d^2 \ln^2 \tau}{2}\right) \binom{\ell}{\ell/2 + r_d} \right.$$
$$\left. + \exp\left(\frac{-d(1-\delta_d)^2 \ln^2 \tau}{2}\right) \binom{\ell}{\ell/2 + r_d + 1} \right\}.$$

We plot the bound given by this lemma in Figure 2. Note how it completely captures the behavior of the degree distribution (barring a slight inaccuracy for larger degrees of the Graph500 graph because we start exceeding the upper bound for d in Lemma 3.2). Theorem 3.1 can be derived from this lemma, as we show here.

PROOF OF THEOREM 3.1. Since $\delta_d = \theta_d - \lfloor \theta_d \rfloor = \theta_d - r_d$, only one of δ_d and $(1 - \delta_d)$ is at most 1/2. In the former case, $\Gamma_d = r_d$ and in the latter case, $\Gamma_d = r_d + 1$. Suppose that $\Gamma_d = r_d$. Then,

$$\exp\left(\frac{-d(1-\delta_d)^2 \ln^2 \tau}{2}\right) \binom{\ell}{\ell/2 + r_d + 1} \leq \exp\left(\frac{-d \ln^2 \tau}{8}\right) \binom{\ell}{\ell/2 + r_d + 1}.$$

Note that this is a small (additive) exponential term in Lemma 3.2. So we just neglect it (and drop the leading constant of $1/\sqrt{2\pi}$) to get a simple approximation. A similar argument works when $\Gamma_d = r_d + 1$. □

In the next section, we prove some preliminary claims which are building blocks in the proof of Lemma 3.2. Then, we give a long intuitive explanation of how we prove Lemma 3.2. Finally, in Section 3.5, we give a complete proof of Lemma 3.2.

3.3. Preliminaries

We will state and prove some simple and known results in our own notation. This will give the reader some understanding about the various *slices* of vertices, and how the degree distribution is related to these slices. Our first claim computes the probability that a single edge insertion creates an outedge for node v. The probability depends only on the slice that v is in.

CLAIM 3.3. *For vertex $v \in S_r$, the probability that a single edge insertion in SKG produces an out-edge at node v is*

$$p_r = \frac{(1 - 4\sigma^2)^{\ell/2}\tau^r}{n}.$$

PROOF. We consider a single edge insertion. What is the probability that this leads to an outedge of v? At every level of the insertion, the edge must go into the half corresponding to the binary representation of v. If the first bit of v is 0, then the edge should drop in the top half at the first level, and this happens with probability $(1/2+\sigma)$. On the other hand, if this bit is 1, then the edge should drop in the bottom half, which happens with probability $(1/2 - \sigma)$. By performing this argument for every level, we get that

$$p_r = \left(\frac{1}{2} + \sigma\right)^{\ell/2+r}\left(\frac{1}{2} - \sigma\right)^{\ell/2-r} = \frac{(1 - 4\sigma^2)^{\ell/2}}{2^\ell} \cdot \left(\frac{1/2+\sigma}{1/2-\sigma}\right)^r = \frac{(1 - 4\sigma^2)^{\ell/2}\tau^r}{n}. \qquad \square$$

Our next lemma bounds the probability that a vertex v at slice r has degree d. Before that, we separately deal with slices where p_r is very large. Essentially, we show that slices where $p_r \geq 1/\sqrt{m}$ can be ignored. This allows for simpler calculations later on.

CLAIM 3.4. *Let R be the set $\{r | p_r \geq 1/\sqrt{m}\}$ and $U = \bigcup_{r \in R} S_r$. The probability that any vertex in U has degree less than $\sqrt{m}/2$ is at most $e^{-\Omega(\sqrt{m})}$.*

PROOF. Consider a fixed $v \in U$. Let X_i be the indicator random variable for the ith edge insertion being incident to v. The X_is are i.i.d. with $\mathbf{E}[X_i] \geq 1/\sqrt{m}$. The out-degree of v is $X = \sum_{i=1}^m X_i$ and $\mathbf{E}[X] \geq \sqrt{m}$. By a multiplicative Chernoff bound (Theorem 4.2 of Motwani and Raghavan [1995]), the probability that $X \leq \sqrt{m}/2$ is at most $e^{-\sqrt{m}/8}$. The proof is completed by taking a union bound over all vertices in U and noting that $ne^{-\sqrt{m}/8} = e^{-\Omega(\sqrt{m})}$. $\square$

We will set $d = o(\sqrt{n})$. Our formula becomes slightly inaccurate when d becomes large, but as our figures show, it is not a major issue in practice. The previous claim implies that the expected number of vertices in U (as defined above) with degree d is vanishingly small. Therefore, we only need to focus on slices where $p_r \leq 1/\sqrt{m}$.

LEMMA 3.5. *Let v be a vertex in slice r. Assume that $p_r \leq 1/\sqrt{m}$ and $d = o(\sqrt{n})$. Then for SKG,*

$$\Pr[\deg(v) = d] = (1 + o(1))\frac{\lambda^d}{d!}\frac{(\tau^r)^d}{\exp(\lambda\tau^r)}.$$

PROOF. The probability that v has outdegree d is $\binom{m}{d}p_r^d(1 - p_r)^{m-d}$. Since $d = o(\sqrt{n})$, we have $\binom{m}{d} = (1 \pm o(1))m^d/d!$. For $x \leq 1/\sqrt{m}$ and $m' \leq m$, we can use the Taylor series approximation, $(1 - x)^{m'} = (1 \pm o(1))e^{-xm'}$. Using Claim 3.3, we get

$$\binom{m}{d}p_r^d(1 - p_r)^{m-d} = (1 \pm o(1))\frac{m^d}{d!}\left(\frac{(1 - 4\sigma^2)^{\ell/2}\tau^r}{n}\right)^d \exp\left(-\frac{(1 - 4\sigma^2)^{\ell/2}\tau^r(m-d)}{n}\right)$$

$$= (1 \pm o(1))\frac{\left(\Delta(1 - 4\sigma^2)^{\ell/2}\right)^d \tau^{rd}}{d!} \exp\left(-\Delta(1 - 4\sigma^2)^{\ell/2}\tau^r\right)$$

$$\times \exp\left(\frac{d(1 - 4\sigma^2)^{\ell/2}\tau^r}{n}\right)$$

$$= (1 \pm o(1))\frac{\lambda^d}{d!}\frac{(\tau^r)^d}{\exp(\lambda\tau^r)}\exp(dp_r).$$

Since $p_r \leq 1/\sqrt{m}$ and $d = o(\sqrt{n})$, $dp_r = o(1)$, completing the proof. □

3.4. Understanding the Degree Distribution

The following is a verbal explanation of our proof strategy and captures the essence of the math.

It will be convenient to think of the parameters having some fixed values. Let $\lambda = 1$ and $\tau = e$. (This can be achieved with a reasonable choice of T, ℓ, Δ.) We begin by looking at the different slices of vertices. Vertices in a fixed r-slice have an identical behavior with respect to the degree distribution. Lemma 3.5 uses elementary probability arguments to argue that the probability that a vertex in slice r has outdegree d is roughly

$$\Pr[\deg(v) = d] = \frac{\exp(dr - e^r)}{d!}. \tag{1}$$

When $r = \Omega(\ln d)$, the numerator will be less than 1, and the overall probability is $O(1/d!)$. Therefore, those slices will not have many (or any) vertices of degree d. If $r = O(\ln d)$, the numerator is $o(d!)$ and the probability is still (approximately) at most $1/d!$. Observe that when r is negative, then this probability is extremely small, even for fairly small values of d. This shows that half of the vertices (in slices where the number of 1's is more than 0's) have extremely small degrees.

It appears that the "sweet spot" is around $r \approx \ln d$. Applying Taylor approximations to appropriate ranges of r, it can be shown that a suitable approximation of the probability of a slice r vertex having degree d is roughly $\exp(-d(r - \ln d)^2)$. We can now show that the SKG degree distribution is bounded *above* by a lognormal tail. Only the vertices in slice $r \approx \ln d$ have a good chance of having degree d. This means that the expected number of vertices of degree d is at most $\binom{\ell}{\ell/2 + \ln d}$. Since the latter is asymptotically normally distributed as a function of $\ln d$, it (approximately) represents a lognormal tail. A similar conclusion was drawn in Kim and Leskovec [2010], though their approach and presentation is very different from ours.

This is where we significantly diverge. The crucial observation is that r is a *discrete* variable, not a continuous one. When $|r - \ln d| \geq 1/3$ (say), the probability of having degree d is at most $\exp(-d/9)$. That is an exponential tail, so we can safely assume that vertices in those slices have no vertices of degree d. Refer to Figure 3. Since $\ln d$ is not necessarily integral, it could be that for *all* values of r, $|r - \ln d| \geq 1/3$. In that case, there are (essentially) no vertices of degree d. For concreteness, suppose $\ln d = 100/3$. Then, regardless of the value of r, $|r - \ln d| \geq 1/3$. And we can immediately bound the fraction of vertices that have this degree by the exponential tail, $\exp(-d/9)$. When $\ln d$ is close to being integral, then for $r = \lfloor \ln d \rceil$, the r-slice (and only this slice) will contain

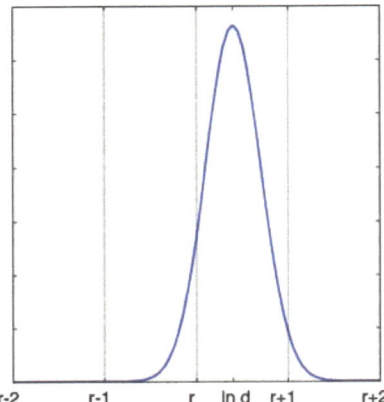

Fig. 3. Probability of nodes of degree d for various slices. The probability that a vertex of slice r has degree d is Gaussian distribution with a peak at $\ln d$. The standard deviation is extremely small. Hence, if $\ln d$ is far from integral, *no slice* will have vertices of degree d.

many vertices of degree d. The quantity $|\ln d - \lfloor \ln d \rceil|$ fluctuates between 0 and $1/2$, leading to the oscillations in the degree distribution.

Let $\Gamma_d = \lfloor \ln d \rceil$ and $\gamma_d = |\Gamma_d - \ln d|$. Putting the arguments above together, we can get a very good estimate of the number of vertices of degree d. This quantity is essentially $\exp(-\gamma_d^2 d)\binom{\ell}{\ell/2+\Gamma_d}$, as stated in Theorem 3.1. A more nuanced argument leads to the bound in Lemma 3.2.

3.5. Proof of Lemma 3.2

We break up the main argument into various claims. The first claim gives an expression for the expected number of vertices of degree d. This sum will appear to be a somewhat complicated sum of binomial coefficients. But, as we later show, we can deduce that most terms in this sum are actually negligible.

CLAIM 3.6. *Define $g(r) = r \ln \tau - \ln(d/\lambda)$. Then, for SKG,*

$$\mathbf{E}[X_d] = \frac{1 \pm o(1)}{\sqrt{2\pi d}} \sum_{r=-\ell/2}^{\ell/2} \exp\left[d(1 + g(r) - e^{g(r)})\right]\binom{\ell}{\ell/2 + r}.$$

PROOF. Using Lemma 3.5 and linearity of expectation, we can derive a formula for $\mathbf{E}[X_d]$. We then apply Stirling's approximation and the fact that $|S_r| = \binom{\ell}{\ell/2+r}$.

$$\mathbf{E}[X_d] = (1 \pm o(1))\frac{\lambda^d}{d!} \sum_{r=-\ell/2}^{\ell/2} \frac{(\tau^r)^d}{\exp(\lambda \tau^r)}|S_r|$$

$$= (1 \pm o(1))\frac{\lambda^d}{d!} \sum_{r=-\ell/2}^{\ell/2} \frac{(\tau^r)^d}{\exp(\lambda \tau^r)}\binom{\ell}{\ell/2 + r}$$

$$= \frac{1 \pm o(1)}{\sqrt{2\pi d}}\left(\frac{e\lambda}{d}\right)^d \sum_{r=-\ell/2}^{\ell/2} \frac{(\tau^r)^d}{\exp(\lambda \tau^r)}\binom{\ell}{\ell/2 + r}.$$

Let us now focus on the quantity

$$\left(\frac{e\lambda}{d}\right)^d \frac{(\tau^r)^d}{\exp(\lambda\tau^r)} = \exp(d + d\ln\lambda + rd\ln\tau - d\ln d - \lambda\tau^r).$$

The term inside the exponent can be written as $d + d(r\ln\tau - \ln d + \ln\lambda) - d(d/\lambda)^{-1}\tau^r$. This is $d(1 + g(r) - e^{g(r)})$. Hence

$$\mathbf{E}[X_d] = \frac{1 \pm o(1)}{\sqrt{2\pi d}} \sum_{r=-\ell/2}^{\ell/2} e^{d(1+g(r)-e^{g(r)})}\binom{\ell}{\ell/2 + r}. \qquad \square$$

The key observation is that among the ℓ terms in the summation of Claim 3.6, few of them are the main contributors. All other terms sum up to a negligible quantity. We deal with this part in the following claim. We crucially use the assumption that $d > (e\ln 2)\ell$. This ensures that the large slices (when $|r|$ is small) do not contribute vertices of degree d.

CLAIM 3.7. *Let R be the set of r such that $|g(r)| \geq 1$. Then, for SKG,*

$$\sum_{r\in R} \exp\left[d(1 + g(r) - e^{g(r)})\right]\binom{\ell}{\ell/2 + r} \leq 1.$$

PROOF. For convenience, define $h(r) = 1 + g(r) - e^{g(r)}$. We will show (shortly) that when $|g(r)| \geq 1$, $h(r) \leq -1/e$. We assume $d > (e\ln 2)\ell$, thus $\exp(d \cdot h(r)) \leq 2^{-\ell}$. Let R be the set of all r such that $|g(r)| \geq 1$. We can easily bound the contribution of the indices in R to our total sum as

$$\sum_{r\in R} e^{dh(r)}\binom{\ell}{\ell/2 + r} \leq 2^{-\ell}\sum_{r\in R}\binom{\ell}{\ell/2 + r} \leq 1.$$

It remains to prove the bound on $h(r)$. Set $\hat{h}(x) = 1 + x - e^x$, so $h(r) = \hat{h}(g(r))$. We have two cases.

—$g(r) \geq 1$: Since $\hat{h}(x)$ is decreasing when $x \geq 1$, $h(r) \leq \hat{h}(1) = -(e-2) \leq -1/e$.
—$g(r) \leq -1$: Since $\hat{h}(x)$ is increasing for $x \leq -1$, $h(r) \leq \hat{h}(-1) = -1/e$. $\qquad\square$

Now for the main technical part. The following claim with the previous ones complete the proof of Lemma 3.2.

CLAIM 3.8. *Define R as in Claim 3.7. Then, for SKG,*

$$\sum_{r\notin R} \exp\left[d(1 + g(r) - e^{g(r)})\right]\binom{\ell}{\ell/2 + r} = (1 \pm o(1)) \cdot \left\{\exp\left(\frac{-d\delta_d^2\ln^2\tau}{2}\right)\binom{\ell}{\ell/2 + r_d}\right.$$

$$\left. + \exp\left(\frac{-d(1-\delta_d)^2\ln^2\tau}{2}\right)\binom{\ell}{\ell/2 + r_d + 1}\right\}.$$

PROOF. Since $|g(r)| < 1$, we can perform an important approximation. Using the expansion $e^x = 1 + x + x^2/2 + \Theta(x^3)$ for $x \in (0, 1)$, we bound

$$h(r) = 1 + g(r) - e^{-g(r)} = -g(r)^2/2 + \Theta(g(r)^3)$$

We request the reader to pause and consider the ramifications of this approximation. The coefficient multiplying the binomial coefficients in the sum is $\exp(-d(g(r))^2)$, which is a Gaussian function of $g(r)$. This is what creates the Gaussian-like behavior of the probability of vertices of degree d among the various slices. We now need to understand when $g(r)$ is close to 0, since the corresponding terms will provide the main contribution to our sum. So for any d, some slices are "picked out" to have expected degree d, whereas others are not. This depends on what the value of $g(r)$ is. Now on, it only requires (many) tedious calculations to get the final result.

What are the different possible values of $g(r)$? We remind the reader that $g(r) = r \ln \tau - \ln(d/\lambda)$. Observe that $r_d = \lfloor \ln(d/\lambda)/ \ln \tau \rfloor$ minimizes $|g(r)|$ subject to $g(r) < 0$ and $r_d + 1$ (which is the corresponding ceiling) minimizes $|g(r)|$ subject to $g(r) \geq 0$. For convenience, denote r_d by r_f (for floor) and $r_d + 1$ by r_c (for ceiling).

Consider some r such that $|g(r)| < 1$. It is either of the form $r = r_c + s$ or $r_f - s$, for integer $s \geq 0$. We will sum up all the terms corresponding to the each set separately. For convenience, denote the former set of values of s's such that $|g(r_c + s)| < 1$ by S_1, and define S_2 with respect to $r_f - s$. This allows us to split the main sum into two parts, which we deal with separately.

Case 1 (the sum over S_1)

$$\sum_{s \in S_1} \exp\left[d(1 + g(r) - e^{g(r)})\right] \binom{\ell}{\ell/2 + r_c + s} = (1 \pm o(1)) \exp\left(\frac{-d(g(r_c)^2)}{2}\right) \binom{\ell}{\ell/2 + r_c}$$

$$+ (1 \pm o(1)) \sum_{\substack{s \in S_1 \\ s \neq 0}} \exp\left(\frac{-d(g(r_c + s)^2)}{2}\right) \binom{\ell}{\ell/2 + r_c + s}$$

We substitute $g(r_c + s) = g(r_c) + s \ln \tau$ into the second part, and show that we can bound this whole summation as an error term. Note that both s and $\ln \tau$ are positive by construction.

$$\sum_{s \in S_1, s \neq 0} \exp\left(-d(g(r_c + s)^2)/2\right) \binom{\ell}{\ell/2 + r_c + s}$$

$$\leq \sum_{s \in S_1, s \neq 0} \exp\left[-d(g(r_c)^2 + s^2(\ln \tau)^2)/2\right] \binom{\ell}{\ell/2 + r_c + s}$$

$$\leq \exp\left(-d(g(r_c)^2)/2\right) \sum_{s > 0} \exp\left(-ds^2(\ln \tau)^2/2\right) \binom{\ell}{\ell/2 + r_c + s}$$

$$= o\left(\exp\left(-d(g(r_c)^2)/2\right) \binom{\ell}{\ell/2 + r_c}\right).$$

For the last inequality, observe that $\binom{\ell}{\ell/2 + r_c + s} \leq \ell^s \binom{\ell}{\ell/2 + r_c}$. Since $d \geq \ell$, the exponential decay of $\exp(\Theta(-ds^2))$ completely kills this summation.

Case 2 (the sum over S_2). Now, we apply an identical argument for $r = r_f - s$. We have $g(r) = g(r_f) - s \ln \tau$. Applying the same calculations as previously mentioned:

$$\sum_{s \in S_2} \exp\left[d(1 + g(r) - e^{g(r)})\right] \binom{\ell}{\ell/2 + r_f + s} = (1 \pm o(1)) \exp\left(-d(g(r_f)^2)/2\right) \binom{\ell}{\ell/2 + r_f}$$

Table IV. Parameters for NSKG

—b = noise parameter $\leq \min((t_1 + t_4)/2, t_2)$
—μ_i = noise at level $i = 1, \ldots, \ell$
—$T_i = \begin{bmatrix} t_1 - \frac{2\mu_i t_1}{t_1 + t_4} & t_2 + \mu_i \\ t_3 + \mu_i & t_4 - \frac{2\mu_i t_4}{t_1 + t_4} \end{bmatrix}$ = noisy generating matrix at level $i = 1, \ldots, \ell$

Adding the bounds from both the cases, we conclude

$$\sum_{r \notin R} \exp\left[d(1 + g(r) - e^{g(r)})\right] \binom{\ell}{\ell/2 + r}$$

$$= (1 \pm o(1)) \cdot \left\{ \exp\left(-dg(r_f)^2/2\right) \binom{\ell}{\ell/2 + r_f} + \exp\left(-dg(r_c)^2/2\right) \binom{\ell}{\ell/2 + r_c} \right\} \quad (2)$$

We showed earlier that $r_f = r_d$ and $r_c = r_d + 1$. We remind the reader that $\theta_d = \ln(d/\lambda)/\ln \tau$, $r_d = \lfloor \theta_d \rfloor$, and $\delta_d = \theta_d - r_d$. Hence $g(r_f) = g(\theta_d) - \delta_d \ln \tau = -\delta_d \ln \tau$. Since $r_c = r_f + 1$, $g(r_c) = \ln \tau + g(r_f) = (1 - \delta_d) \ln \tau$. We substitute in (2) to complete the proof. □

4. ENHANCING SKG WITH NOISE: NSKG

Let us now focus on a noisy version of SKG that removes the fluctuations in the degree distribution. We will refer to our proposed noisy SKG model as NSKG. The idea is quite simple. For each level $i \leq \ell$, define a new matrix T_i in such a way that the expectation of T_i is just T. At level i in the edge insertion, we use the matrix T_i to choose the appropriate quadrant.

Here is a formal description. For convenience, we will assume that T is symmetric. It is fairly easy to generalize to general T. Let b be our noise parameter such that $b \leq \min((t_1 + t_4)/2, t_2)$. For level i, choose μ_i to be a uniform random number in the range $[-b, +b]$. Set T_i to be

$$T_i = \begin{bmatrix} t_1 - \frac{2\mu_i t_1}{t_1 + t_4} & t_2 + \mu_i \\ t_3 + \mu_i & t_4 - \frac{2\mu_i t_4}{t_1 + t_4} \end{bmatrix}.$$

Note that T_i is symmetric, its entries sum to 1, and all entries are positive. This is by no means the only model of noise, but it is certainly convenient for analysis. Each level involves only one random number μ_i, which changes all the entries of T in a linear fashion. Hence, we only need ℓ random numbers in total. For convenience, we list out the noise parameters of NSKG in Table IV.

In Figures 1, 4(a), and 4(b), we show the effects of noise. Observe how even a noise parameter as small as 0.05 (which is extremely small compared to the matrix values) significantly reduces the magnitude of oscillations. A noise of 0.1 almost removes the oscillations. (Even this noise is very small, since the standard deviation of this noise parameter is at most 0.06.) Our proposed method of adding noise dampens the undesirable exponential tail behavior of SKG, leading to a monotonic degree distribution.

4.1. Why Does Noise Help?

Before we state our formal theorem, let us set some asymptotic notation that will allow for a more readable theorem. We will use the $O(\cdot)$ notation to suppress constant factors, where (for notational convenience) these constants *may* depend on the constants in the matrix T. As before, $o(1)$ is a quantity that goes to zero as ℓ grows.

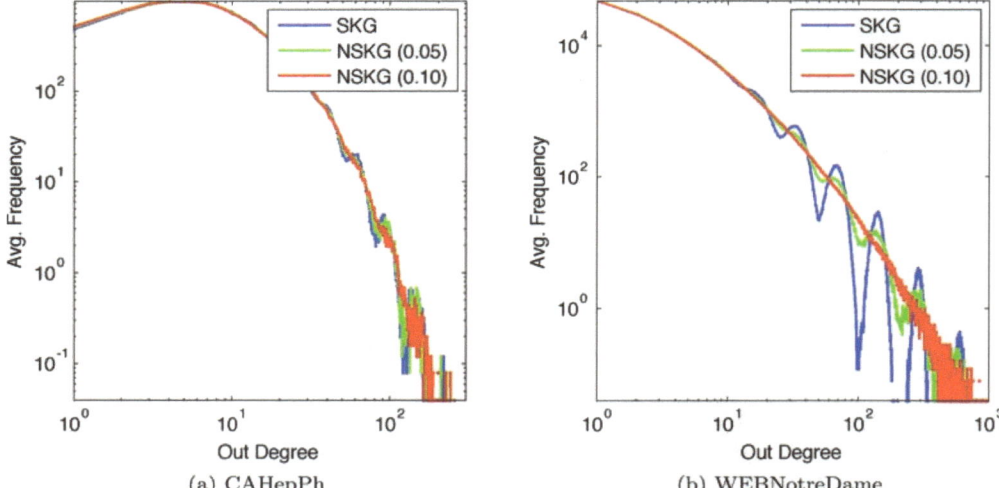

Fig. 4. The figures show the degree distribution of standard SKG and NSKG as the averages of 25 instances. Notice how effectively a noise of 0.1 straightens the degree distribution.

Our formal theorem says that when the noise is "large enough," we can show that the degree distribution has at least a lognormal tail on average. This is a significant change from SKG, where many degrees are below an exponential tail.

THEOREM 4.1. *Let noise b be set to $c/\sqrt{\ell}$ for positive c, such that $c/\sqrt{\ell} < \min((t_1 + t_4)/2, t_2)$. Then, the expected degree distribution for NSKG is bounded below by a lognormal. Formally, when $\Gamma_d \leq \ell/2$ and $d \leq \sqrt{n}$,*

$$\mathbf{E}[X_d] \geq \frac{v(c)}{d}\binom{\ell}{\ell/2 + \Gamma_d}.$$

Here $v(c)$ is some positive function of c. (This is independent of ℓ, so for constant c, $v(c)$ is a positive constant.)

This bound tells us that as ℓ increases, we need *less* noise to get a lognormal tail. From a Graph 500 perspective, if we determine (through experimentation) that for some small ℓ a certain amount of noise suffices, the *same* amount of noise is certainly enough for *larger* ℓ.

We now provide a verbal description of the main ideas. Let us assume that $\lambda = 1$ and $\tau = e$, as before. We focus our attention on a vertex v of slice r, and wish to compute the probability that it has degree d. Note the two sources of randomness: one coming from the choice of the noisy SKG matrices, and the second from the actual graph generation. We associate a *bias parameter* ρ_v with every vertex v. This can be thought of as some measure of how far the degree behavior of v deviates from its noiseless version. Actually, it is the random variable $\ln \rho_v$ that we are interested in. Intuitively, this can just be thought of as a Gaussian random variable with mean zero. The *distribution* of ρ_v is identical for all vertices in slice r. (Though it does not matter for our purposes, for a given instantiation of the noisy SKG matrices, vertices in the same slice can have different biases.)

We approximate the probability that v has degree d by (refer to Claim 4.11)

$$\Pr[\deg(v) = d] = \exp\left(dr + d\ln \rho_v - \rho_v e^r\right)/d!.$$

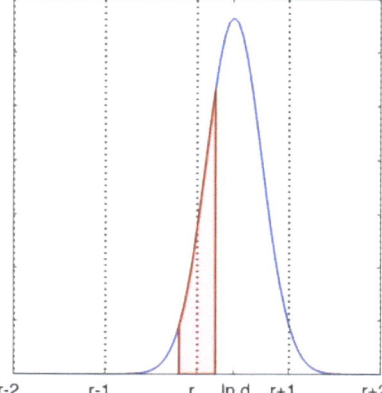

Fig. 5. The effect of noise. The underlying Gaussian curve is the same as one in Figure 3. Adding noise can be thought of as an average over the Gaussian. So the probability that a vertex from slice r has degree is the area of the shaded region.

After some simplifications, this is roughly equal to $\exp(-d(r - \ln d - \ln \rho_v)^2)$. The additional $\ln \rho_v$ will act as a smoothing term. Observe that even if $\ln d$ has a large fractional part, we could still get vertices of degree d. Suppose $\ln d = 10.5$, but $\ln \rho_v$ happened to be close to 0.5. Then, vertices in slice $\lfloor \ln d \rceil$ would have degree d with some nontrivial probability. Contrast this with regular SKG, where there is almost no chance that degree d vertices exist.

Think of the probability as $\exp(d(r - \ln d - X)^2)$, where X is a random variable. The expected probability will be an average over the distribution of X. Intuitively, instead of the probability just being $\exp(d(r - \ln d)^2)$ (in the case of SKG), it is now the *average* value over some interval. If the standard deviation of X is sufficiently large, even though $\exp(d(r - \ln d)^2)$ is small, the average of $\exp(d(r - \ln d - X)^2)$ can be large. Refer to Figure 5.

We know that X is a Gaussian random variable (with some standard deviation σ). So we can formally express the (expected) probability that v has degree d as an integral,

$$P(\deg(v) = d \mid \tau = e, \ \lambda = 1) = \int_{-\infty}^{+\infty} \exp(d(r - \ln d - X)^2) \cdot e^{-X^2/2\sigma^2} dX.$$

This definite integral can be evaluated exactly (since it is just a Gaussian). Intuitively, this is roughly the average value of $\exp(d(r - \ln d - X)^2)$, where X ranges from $-\sigma$ to $+\sigma$. Suppose $\sigma > 1$. Since r ranges over the integers, there is always *some* r such that $|r - \ln d| < 1$. For this value of r, the average of $\exp(d(r - \ln d - X)^2)$ over the range $X \in [-1, +1]$ will have a reasonably large value. This ensures that (in expectation) many vertices in this slice r have degree d. This can be shown for all degrees d, and we can prove that the degree distribution is at least lognormal.

This is an intuitive sketch of the proof. The random variable $\ln \rho_v$ is not exactly Gaussian, and hence we have to account for errors in such an approximation. We do not finally get a definite integral that can be evaluated exactly, but we can give good bounds for its value.

4.2. Preliminaries for Analysis

There are many new parameters we need to introduce for our NSKG analysis. Each of these quantities is a random variable that depends on the choice of the matrices T_i. We list them here.

—$\sigma_i = t_1 - \frac{2\mu_i t_1}{t_1 + t_4} + t_2 + \mu_i - 0.5 = \sigma + \mu_i(1 - \frac{2t_1}{t_1 + t_4})$.

—$\alpha_i = (1/2 + \sigma_i)/(1/2 + \sigma)$. It will be convenient to express this in terms of μ_i, replacing the dependence on σ_i.

$$\alpha_i = (1/2 + \sigma_i)/(t_1 + t_2) = 1 - \mu_i \frac{(t_1 - t_4)}{(t_1 + t_2)(t_1 + t_4)}$$

—$\beta_i = (1/2 - \sigma_i)/(1/2 - \sigma)$. Performing a calculation similar to this one.

$$\beta_i = (1/2 - \sigma_i)/(t_3 + t_4) = 1 + \mu_i \frac{(t_1 - t_4)}{(t_3 + t_4)(t_1 + t_4)}$$

—b_α, b_β: We set

$$b_\alpha = \frac{b(t_1 - t_4)}{(t_1 + t_2)(t_1 + t_4)} = \frac{4b\sigma}{(1 + 2\sigma)(t_1 + t_4)}$$

Similarly,

$$b_\beta = \frac{b(t_1 - t_4)}{(t_3 + t_4)(t_1 + t_4)} = \frac{4b\sigma}{(1 - 2\sigma)(t_1 + t_4)}$$

Hence, α_i is distributed uniformly at random in $[1 - b_\alpha, 1 + b_\alpha]$, and β_i is uniformly random in $[1 - b_\beta, 1 + b_\beta]$. Note that $b_\alpha, b_\beta = \Theta(c/\sqrt{\ell})$.

—ρ_v: Let v be represented as a bit vector $(z_1, \ldots, z_k)$. The *bias* for v is $\rho_v = \prod_{i:z_i=0} \alpha_i \prod_{i:z_i=1} \beta_i$. We set $\lambda_v = \lambda \rho_v$.

4.3. The Behavior of $\ln \rho_v$

We need to bound the behavior of $\ln \rho_v$, which is $\sum_{i:z_i=0} \ln \alpha_i + \sum_{i:z_i=1} \ln \beta_i$. Observe that this is a sum of independent random variables. By the Central Limit Theorem, we expect $\ln \rho_v$ to be distributed as a Gaussian, but we still need to investigate the variance of this distribution. Approximately (since b_α and b_β are small), $\ln \alpha_i$ is uniformly random in $[-b_\alpha, b_\alpha]$, so the variance of $\ln \alpha_i$ is $\Theta(b_\alpha^2) = \Theta(1/\ell)$. A similar statement holds for $\ln \beta_i$, and we bound the variance of $\ln \rho_v$ by $\Theta(1)$. So the probability density function (pdf) of $\ln \rho_v$ is roughly concentrated in a constant-sized interval of size 1 (around 0). This is what we will formally show in this section. We will need a pointwise convergence guarantee for the pdf of $\ln \rho_v$. Throughout this section, we will use various functions of the form $v_1(c), v_2(c), \ldots$. These are strictly positive constant functions of c (for $c > 0$), and are a convenient way of tracking dependences on c. The reader should interpret $v_a(c)$ to be some constant that depends on c (and T and Δ, which are fixed), but is independent of ℓ. The main lemma of this section is the following.

LEMMA 4.2. *Set $\hat{\tau} = \max(\ln \tau, 2)$. Let $f_v(x)$ be the pdf of $\ln \rho_v$. For $|x| \leq \hat{\tau}$, $f_v(x) \geq v_1(c)$.*

We will first prove Lemma 4.2 as a direct result of two claims stated here. Then we will prove these claims in the subsequent subsections. The first claim, the more technical of the two, shows that $\ln \rho_v$ has a sufficiently large probability of attaining a constant value.

CLAIM 4.3. *There exists a constant $C > \hat{\tau}$, such that the probability that $\ln \rho_v$ lies in $[\hat{\tau}, C]$ is at least $v_2(c)$ and that of lying in $[-C, -\hat{\tau}]$ is also at least $v_2(c)$.*

The next claim will be a consequence of the unimodularity of $f_v(x)$.

CLAIM 4.4. *For any $x \in [x_1, x_2]$, $f_v(x) \geq \min(f_v(x_1), f_v(x_2))$.*

Now for the proof of Lemma 4.2.

PROOF OF LEMMA 4.2. By Claim 4.3, the probability that $\ln \rho_v$ lies in $I := [-C, -\hat{\tau}]$ is at least $\nu_2(c)$. Therefore, $(C - \hat{\tau}) \max_{x \in I} f_v(x) \geq \nu_2(c)$. Suppose the maximum is achieved at x_1. This means that there exists $x_1 \in [-C, -\hat{\tau}]$, $f_v(x_1) = \Omega(\nu_2(c))$. Similarly, there exists some $x_2 \in [\hat{\tau}, C]$ such that $f_v(x_2) = \Omega(\nu_2(c))$. Observe that for any x such that $|x| \leq \hat{\tau}$, $x \in [x_1, x_2]$. By Claim 4.4, for any such x, $f_v(x) = \Omega(\nu_2(c))$. Therefore, we can bound $f_v(x) \geq \nu_1(c)$, for some positive function ν_1. □

4.3.1. Proving Claim 4.3. We begin with notational setup. We fix some vertex v. For convenience, define the variables $\widehat{\alpha}_i$ (for all $i \leq \ell$). If $z_i = 0$, set $\widehat{\alpha}_i = \alpha_i$ and $\widehat{\alpha}_i = \beta_i$ otherwise. We can write $\ln \rho_v = \sum_i \ln \widehat{\alpha}_i$. The random variable $\widehat{\alpha}_i$ is uniform in $[1 - b_i, 1 + b_i]$, where b_i is either b_α or b_β appropriately. Set the zero mean random variable $X_i = \ln \widehat{\alpha}_i - \mathbf{E}[\ln \widehat{\alpha}_i]$. We have the following series of facts.

CLAIM 4.5.

—*The pdf of $\ln \widehat{\alpha}_i$, denoted by $h_i(x)$, is given as follows. For $x \in [\ln(1 - b_i), \ln(1 + b_i)]$, $h_i(x) = e^x/2b_i$, and zero otherwise.*
—$|\mathbf{E}[\ln \widehat{\alpha}_i]| = O(c^2/\ell)$, $\mathbf{E}[X_i^2] = \Theta(c^2/\ell)$, *and* $\mathbf{E}[|X_i|^3] = O(c\mathbf{E}[X_i^2]/\sqrt{\ell})$.

PROOF. The pdf of $\widehat{\alpha}_i$ is $h_\alpha(x) = 1/2b_\alpha$ for $x \in [1 - b_\alpha, 1 + b_\alpha]$ and zero otherwise. For any monotone function $F(x)$, the pdf of $F(\widehat{\alpha}_i)$ is given by $|dF^{-1}(x)/dx|h(x)$. Setting F as the function $\ln$, the pdf of $\ln \alpha_i$, $h_i(x)$, is given by $e^x/2b_\alpha$ for $x \in [\ln(1 - b_\alpha), \ln(1 + b_\alpha)]$ and zero otherwise.

$$\mathbf{E}[\ln \widehat{\alpha}_i] = \int_{\ln(1-b_i)}^{\ln(1+b_i)} x h_i(x)dx = (2b_i)^{-1} \int_{\ln(1-b_i)}^{\ln(1+b_i)} x e^x dx.$$

Using integration by parts,

$$
\begin{aligned}
\int_{\ln(1-b_i)}^{\ln(1+b_i)} x e^x dx &= [x e^x] \Big|_{\ln(1-b_i)}^{\ln(1+b_i)} - \int_{\ln(1-b_i)}^{\ln(1+b_i)} e^x dx \\
&= [(1 + b_i)\ln(1 + b_i) - (1 - b_i)\ln(1 - b_i)] - [(1 + b_i) - (1 - b_i)] \\
&= b_i \ln\left(1 - b_i^2\right) + \ln(1 + b_i) - \ln(1 - b_i) - 2b_i.
\end{aligned}
$$

Taking absolute values,

$$\left| \int_{\ln(1-b_i)}^{\ln(1+b_i)} x e^x dx \right| \leq |b_i \ln\left(1 - b_i^2\right)| + |\ln(1 + b_i) - \ln(1 - b_i) - 2b_i|.$$

The first term is at most $2b_i^3$. For the second term, we need a finer Taylor approximation.

$$\ln(1 + b_i) - \ln(1 - b_i) - 2b_i \leq \left(b_i - b_i^2/2 + b_i^3\right) - \left(-b_i - b_i^2/2\right) - 2b_i \leq b_i^3$$
$$\ln(1 + b_i) - \ln(1 - b_i) - 2b_i \geq \left(b_i - b_i^2/2\right) - \left(-b_i - b_i^2/2 - b_i^3\right) - 2b_i \geq -b_i^3.$$

All in all, $|\mathbf{E}[\ln \widehat{\alpha}_i]| \leq O(b_i^2) = O(c^2/\ell)$.

$$\mathbf{E}[X_i^2] = \mathbf{E}[(\ln \widehat{\alpha}_i)^2] - (\mathbf{E}[\ln \widehat{\alpha}_i])^2$$
$$\mathbf{E}[(\ln \widehat{\alpha}_i)^2] = (2b_i)^{-1} \int_{\ln(1-b_i)}^{\ln(1+b_i)} x^2 e^x dx.$$

To get an upper bound for this term, we use the following inequalities: $\ln(1 + b_\alpha) \leq 2b_\alpha$, $\ln(1 - b_\alpha) \geq -2b_\alpha$, $e^x \leq e$. That gives $\mathbf{E}[(\ln \widehat{\alpha}_i)^2] \leq e(2b_i)^{-1} \int_{-2b_i}^{2b_i} x^2 dx = O(b_i^2)$. For a lower bound, we use: $\ln(1 + b_\alpha) \geq b_\alpha/2$, $\ln(1 - b_\alpha) \leq -b_\alpha/2$, $e^x \geq 1/e$. Hence,

$\mathbf{E}[(\ln \widehat{\alpha}_i)^2] \geq (2eb_i)^{-1} \int_{-b_i/2}^{b_i/2} x^2 dx = \Omega(b_i^2)$. Note that $(\mathbf{E}[\ln \widehat{\alpha}_i])^2 \leq b_i^4$, which is much small than b_i^2 for sufficiently small b_i. We conclude that $\mathbf{E}[X_i^2] = \Theta(b_i^2) = \Theta(c^2/\ell)$.

For the final bound, we use a trivial estimate. We have $\mathbf{E}[|X_i|^3] \leq \max(|X_i|)\mathbf{E}[X_i^2] \leq 2b_i\mathbf{E}[X_i^2]$. $\square$

We now state the Berry-Esseen Theorem [Berry 1941; Esseen 1942], a crucial ingredient of our proof. This theorem bounds the convergence rate of a sum of independent random variables to a Gaussian.

THEOREM 4.6 [BERRY-ESSEEN]. *Let $X_1, X_2, \ldots, X_\ell$ be independent random variables with $\mathbf{E}[X_i] = 0$, $\mathbf{E}[X_i^2] = \xi_i^2$, and $\mathbf{E}[|X_i|^3] = \iota_i < \infty$. Let S be the sum $\sum_i X_i / \sqrt{\sum_i \xi_i^2}$. Let $F(x)$ denote the cumulative distribution function (cdf) of S and $\Phi(x)$ be the cdf of the standard normal (the pdf is $(2\pi)^{-1/2}e^{-x^2/2}$). Then, for an absolute constant $C_1 > 0$,*

$$\sup_x |F(x) - \Phi(x)| \leq C_1 \left(\sum_i \xi_i^2 \right)^{-3/2} \sum_i \iota_i.$$

PROOF OF CLAIM 4.3. We set $X = \sum_i X_i = (\ln \rho_v - \mathbf{E}[\ln \rho_v]) / \sqrt{\sum_i \mathbf{E}[X_i^2]}$. By Claim 4.5, $|\mathbf{E}[\ln \rho_v]| = |\sum_i \mathbf{E}[\ln \widehat{\alpha}_i]| \leq \sum_i |\mathbf{E}[\ln \widehat{\alpha}_i]| = O(c^2)$ and $\sum_i \mathbf{E}[X_i^2] = \Theta(c^2)$. Note that X is just an increasing linear function of $\ln \rho_v$. Set function $r(x) = (x - \mathbf{E}[\ln \rho_v]) / \sqrt{\sum_i \mathbf{E}[X_i^2]}$, so $X = r(\ln \rho_v)$. For any interval $I = [x_1, x_2]$, $\Pr[\ln \rho_v \in I] = \Pr[X \in [r(x_1), r(x_2)]]$. Since $|r(\widehat{\tau})|$ is some constant function of c, we can find a constant C such the $r(C)$ is strictly larger than $|r(\widehat{\tau})|$. Setting $y_1 = r(\widehat{\tau})$, $y_2 = r(C)$ and using the notation from Theorem 4.6,

$$
\begin{aligned}
\Pr[X \in [y_1, y_2]] &= F(y_2) - F(y_1) = \Phi(y_2) - \Phi(y_1) + (F(y_2) - \Phi(y_2)) + (\Phi(y_1) - F(y_1)) \\
&\geq \Phi(y_2) - \Phi(y_1) - |F(y_2) - \Phi(y_2)| - |F(y_1) - \Phi(y_1)|.
\end{aligned}
$$

Since $y_1 < y_2$ and are constant functions of c, $\Phi(y_2) - \Phi(y_1) \geq v_3(c)$. By the Berry-Esseen theorem (Theorem 4.6), $|F(x_2) - \Phi(x_2)| + |F(x_1) - \Phi(x_1)| \leq 2C_1(\sum_i \xi_i^2)^{-3/2} \sum_i \iota_i$. By Claim 4.5 $\iota_i = O(c\xi_i^2/\sqrt{\ell})$ and $\sum_i \xi_i^2 = \Theta(c^2)$. So the Berry-Esseen bound is at most $2C_1 c(\sum_i \ell\xi_i^2)^{-1/2} = O(1/\sqrt{\ell})$. By setting C to be a large enough constant, we can ensure that $\Phi(y_2) - \Phi(y_1) > 2C_1 c(\sum_i \ell\xi_i^2)^{-1/2}$.

We deduce that $\Pr[X \in [x_1, x_2]] \geq v_2(c)$, for some positive function v_2. A similar proof holds for $[-C, -\widehat{\tau}]$. $\square$

4.3.2. Proving Claim 4.4. We state some technical definitions and results about convolutions of unimodal functions.

Definition 4.7. A pdf $f(x)$ is *unimodal* if there exists an $a \in \mathbb{R}$ such that f is nondecreasing on $(-\infty, a)$ and nonincreasing on (a, ∞).

A pdf $f(x)$ is *log-concave* if $Q := \{x : f(x) > 0\}$ is an interval and $\ln f(x)$ is a concave function (on the interval Q).

A theorem of Ibragimov [1956] gives some convolution properties of unimodal log-concave functions.

THEOREM 4.8 [IBRAGIMOV]. *Let $f(x)$ be a unimodal log-concave pdf and $g(x)$ be a unimodal pdf. The convolution $f * g$ is also unimodal.*

CLAIM 4.9. *The pdf $f_v(x)$ is unimodal.*

PROOF. We have $\ln \rho_v = \sum_i \ln \widehat{\alpha}_i$. By Claim 4.5, the pdf of $\ln \widehat{\alpha}_i$ is $h_i(x) = e^x/2b_i$. Note that $h_i(x)$ is unimodal. Furthermore, $\ln h_i(x) = x - \ln 2b_i$, which is concave. Since $\ln \rho_v$ is the sum of independent random variables, the pdf $f_v(x)$ is the convolution of the individual pdfs. Repeated applications of Ibragimov's theorem (Theorem 4.8) tells us that $f_v(x)$ is unimodal. □

PROOF OF CLAIM 4.4. By the unimodality of f_v, f_v is either nondecreasing, nonincreasing, or nondecreasing and then nonincreasing in the interval $[x_1, x_2]$. Regardless of which case, for any $y \in [x_1, x_2]$, $f(y) \geq \min(f(x_1), f(x_2))$. □

4.4. Basic Claims for NSKG

We now reprove some of the basic claims for NSKG. Note that when we look at $\mathbf{E}[X_d]$, the expectation is over both the randomness in T and the edge insertions. We use $\mathbf{T}$ to denote the set of matrices $T_1, T_2, \ldots, T_\ell$. Conditioning on $\mathbf{T}$ simply means conditioning on a fixed choice of the noise.

CLAIM 4.10. *Let vertex* $v \in S_r$. *Choose the noise for NSKG at random, and let* $\hat{p}_v$ *be the probability (conditioned on* $\mathbf{T}$*) that a single edge insertion produces an out-edge at* v. *(Note that* $\hat{p}_v$ *is itself a random variable, where the dependence on* $\mathbf{T}$ *is given by* ρ_v*)*

$$\hat{p}_v = \frac{(1 - 4\sigma^2)^{\ell/2} \tau^r \rho_v}{n}.$$

PROOF. This is identical to the proof of Claim 3.3. Consider a single edge insertion. For an edge insertion to be incident to v, the edge must go into the half corresponding to the binary representation of v. If the ith bit of v is 0, then the edge should drop in the top half at this level, and this happens with probability $(1/2 + \sigma_i)$. On the other hand, if this bit is 1, then the edge should drop in the bottom half, which happens with probability $(1/2 - \sigma_i)$. Let the bit representation of v be $(z_1, z_2, \ldots, z_\ell)$. Then,

$$
\begin{aligned}
\hat{p}_v &= \prod_{i:z_i=0} \left(\frac{1}{2} + \sigma_i\right) \prod_{i:z_i=1} \left(\frac{1}{2} - \sigma_i\right) \\
&= \prod_{i:z_i=0} \alpha_i \left(\frac{1}{2} + \sigma\right) \prod_{i:z_i=1} \beta_i \left(\frac{1}{2} - \sigma\right) \\
&= \rho_v \left(\frac{1}{2} + \sigma\right)^{\ell/2+r} \left(\frac{1}{2} - \sigma\right)^{\ell/2-r} \\
&= \frac{\rho_v (1 - 4\sigma^2)^{\ell/2}}{2^\ell} \cdot \left(\frac{1/2 + \sigma}{1/2 - \sigma}\right)^r = \frac{(1 - 4\sigma^2)^{\ell/2} \tau^r \rho_v}{n}. \quad \square
\end{aligned}
$$

As before, we will assume that $\hat{p}_v = o(1/\sqrt{m})$ and $d = o(\sqrt{n})$. Even though $\hat{p}_v$ is a random variable, the probability that it is larger than $1/\sqrt{m}$ can be neglected. (This was discussed in more detail before Lemma 3.5). We stress that in the following, the probability that v has outdegree d is itself a random variable.

CLAIM 4.11. *Let* v *be a vertex in slice* r, $d = o(\sqrt{n})$, *and* $\hat{p}_v = o(1/\sqrt{m})$. *Then, for NSKG, we have*

$$\Pr[\deg(v) = d | \mathbf{T}] = (1 \pm o(1)) \frac{(\lambda_v)^d}{d!} \cdot \frac{(\tau^r)^d}{\exp(\lambda_v \tau^r)}.$$

PROOF. We follow the proof of Lemma 3.5. We approximate $\binom{m}{d}$ by $m^d/d!$ and $(1-x)^{m-d}$ by e^{-xm}, for $x = o(1/\sqrt{m})$ and $d = o(\sqrt{n})$. This approximation is performed in the first step. We remind the reader that $\lambda_v = \lambda \rho_v$. By Claim 4.10 and those approximations,

$$\binom{m}{d}\hat{p}_v^d(1-\hat{p}_v)^{m-d} = \binom{m}{d}\left(\frac{(1-4\sigma^2)^{\ell/2}\tau^r\rho_v}{n}\right)^d\left(1-\frac{(1-4\sigma^2)^{\ell/2}\tau^r\rho_v}{n}\right)^{m-d}$$

$$= (1\pm o(1))\frac{m^d}{d!}\times\left(\frac{(1-4\sigma^2)^{\ell/2}\tau^r\rho_v}{n}\right)^d$$

$$\cdot \exp\left(-\frac{(1-4\sigma^2)^{\ell/2}\tau^r\rho_v m}{n}\right)$$

$$= (1\pm o(1))\frac{\left[\Delta(1-4\sigma^2)^{\ell/2}\rho_v\right]^d\tau^{rd}}{d!}\exp\left(-\Delta(1-4\sigma^2)^{\ell/2}\rho_v\tau^r\right)$$

$$= (1\pm o(1))\frac{(\lambda\rho_v)^d}{d!}\cdot\frac{(\tau^r)^d}{\exp(\lambda\rho_v\tau^r)}. \qquad \square$$

4.5. Bounds for Degree Distribution

We complete the proof of Theorem 4.1. We break it down into some smaller claims. By and large, the flow of the proof is similar to that for the standard SKG. The main difference comes because the probabilities discussed in Claim 4.11 are random variables depending on the noise. The following claim is fairly straightforward, given the previous analysis of standard SKG. This is where we apply the Taylor approximations to show the Gaussian behavior depicted in Figure 3.

CLAIM 4.12. *Consider some setting of the NSKG noise. Define $g_v(r) = r\ln\tau - \ln(d/\lambda_v)$. The expected number of vertices of degree d conditioned on $\mathbf{T}$ is*

$$\mathbf{E}[X_d|\mathbf{T}] = \frac{1\pm o(1)}{\sqrt{2\pi d}}\sum_{r=-\ell/2}^{\ell/2}\sum_{v\in S_r}\exp\left[-dg_v(r)^2/2\right].$$

PROOF. By fixing some $\mathbf{T}$, the λ_vs are fixed. We use Claim 4.11, linearity of expectation, and Stirling's approximation in the following.

$$\mathbf{E}_G[X_d] = \sum_{r=-\ell/2}^{\ell/2}\sum_{v\in S_r}(1\pm o(1))\frac{\lambda_v^d}{d!}\frac{(\tau^r)^d}{\exp(\lambda_v\tau^r)}$$

$$= \frac{1\pm o(1)}{\sqrt{2\pi d}}\sum_{r=-\ell/2}^{\ell/2}\sum_{v\in S_r}\left(\frac{e\lambda_v}{d}\right)^d\frac{(\tau^r)^d}{\exp(\lambda_v\tau^r)}.$$

Choose a $v\in S_r$.

$$\left(\frac{e\lambda_v}{d}\right)^d\frac{(\tau^r)^d}{\exp(\lambda_v\tau^r)} = \exp\left(d+d\ln\lambda_v+rd\ln\tau-d\ln d-\lambda_v\tau^r\right).$$

Define $f_v(r) = rd\ln\tau - \lambda_v\tau^r - d\ln d + d\ln\lambda_v + d$, where r is an integer. We have $r = (\ln d - \ln\lambda_v + g_v(r))/\ln\tau$.

$$f_v(r) = d\ln d - d\ln\lambda_v + dg_v(r) - e^{g_v(r)}d - d\ln d + d\ln\lambda_v + d$$

$$= d\left(1+g_v(r)-e^{g_v(r)}\right).$$

If $|g_v(r)| < 1$, then we can approximate $f_v(r) = -d[g_v(r)^2/2 + \Theta(g_v(r)^3)]$, and get $\exp(f_v(r)) = (1\pm o(1))\exp(-dg_v(r)^2/2)$. This is analogous to the beginning of the proof of Claim 3.8. Suppose $|g_v(r)| \geq 1$. Then, arguing as in the proof of Claim 3.7, we deduce that $\exp(f_v(r)) \leq 2^{-\ell}$. The sum of all these terms over v is just a lower order term. So, we can substitute this by $\exp(-dg_v(r)^2/2)$. Hence, we can bound

$$\mathbf{E}[X_d | \mathbf{T}] = \frac{1 \pm o(1)}{\sqrt{2\pi d}} \sum_{r=-\ell/2}^{\ell/2} \sum_{v \in S_r} \exp\left[-dg_v(r)^2/2\right]. \quad \square$$

We now reach the main challenge of this proof. The quantity $\mathbf{E}[\exp(-dg_v(r)^2/2)]$ is evaluated by averaging over all noise. Note that the actual graph has no effect on this quantity.

Lemma 4.13. *Consider* $r = \Gamma_d = \lceil \theta_d \rceil$.

$$\mathbf{E}[\exp\left(-dg_v(r)^2/2\right)] \geq \frac{v_4(c)}{\sqrt{d}}$$

Proof. Define $\xi_{r,d} = (r - \theta_d) \ln \tau$. Since $\theta_d = \ln(d/\lambda)/\ln \tau$,

$$g_v(r) = r \ln \tau - \ln(d/\lambda_v) = r \ln \tau - \ln(d/\lambda) + \ln \rho_v = \xi_{r,d} + \ln \rho_v.$$

Hence,

$$\mathbf{E}\left[\exp\left(-dg_v(r)^2/2\right)\right] = \mathbf{E}\left[\exp\left[-d(\ln \rho_v + \xi_{r,d})^2/2\right]\right]$$

Since we set $r = \lceil \theta_d \rceil$, $|\xi_{r,d}| \leq (\ln \tau)/2$. Let us now evaluate the expectation. The pdf of $\ln \rho_v$ is denoted by f_v. The expectation is given by an integral. To distinguish the d referring to degree, and the d referring to the infinitesimal, we shall use (d) in parenthesis for the infinitesimal. We hope this slight abuse of notation will not create a problem, since our integrals are not too confusing. By Lemma 4.2, $f_v(x) \geq v_1(c)$ for $|x| \leq \hat{\tau}$.

$$
\begin{aligned}
\mathbf{E}\left[\exp\left(-dg_v(r)^2/2\right)\right] &= \int_{-\infty}^{+\infty} \exp\left[-d(x + \xi_{r,d})^2/2\right] f_v(x)(dx) \\
&\geq v_1(c) \int_{-\hat{\tau}}^{\hat{\tau}} \exp\left[-d(x + \xi_{r,d})^2/2\right](dx) \\
&= v_1(c) \int_{-\hat{\tau}+\xi_{r,d}}^{\hat{\tau}+\xi_{r,d}} \exp[-dx^2/2](dx) \\
&= v_1(c)\left[\int_{-\infty}^{+\infty} \exp[-dx^2/2](dx) - \int_{\hat{\tau}+\xi_{r,d}}^{+\infty} \exp[-dx^2/2](dx)\right. \\
&\qquad \left. - \int_{-\infty}^{-\hat{\tau}+\xi_{r,d}} \exp[-dx^2/2]\right](dx)
\end{aligned}
$$

We have $|\xi_{r,d}| \leq (\ln \tau)/2$ and $\hat{\tau} = \max(2, \ln \tau)$. Hence, $\hat{\tau} + \xi_{r,d} \geq 1$ and $-\hat{\tau} + \xi_{r,d} \leq -1$.

$$
\begin{aligned}
\mathbf{E}\left[\exp\left(-dg_v(r)^2/2\right)\right] &\geq v_1(c)\left[\int_{-\infty}^{+\infty} \exp[-dx^2/2](dx) - \int_{1}^{+\infty} \exp[-dx^2/2](dx)\right. \\
&\qquad \left. - \int_{-\infty}^{-1} \exp\left[-dx^2/2\right]\right](dx) \\
&= (v_1(c)/\sqrt{d})\left[\int_{-\infty}^{+\infty} e^{-x^2/2}dx - 2\int_{\sqrt{d}}^{+\infty} e^{-x^2/2}dx\right]
\end{aligned}
$$

The first integral is just $\sqrt{2\pi}$. The second is a tail probability of the standard Gaussian, bounded by $\int_{y}^{+\infty} e^{-x^2/2}dx < e^{-y^2/2}/y$ (Lemma 2, pg. 175 of Feller [1968]). The second

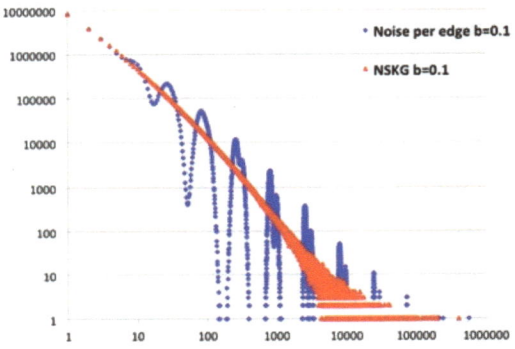

Fig. 6. Comparison of degree distribution of graphs generated by NSKG and by adding noise per edge for Graph500 parameters and $\ell = 26$.

term is at most $2e^{-d^2/2}/\sqrt{d} < \sqrt{\pi}$ (for sufficiently large d). Therefore, we can set function $v_4(c)$ such that $\mathbf{E}[\exp(-dg_v(r)^2/2)] \geq v_4(c)/\sqrt{d}$. □

PROOF OF THEOREM 4.1. This is a direct consequence of the previous claims. Set $r = \Gamma_d$. By Claim 4.12 and linearity of expectation, $\mathbf{E}[X_d] = \mathbf{E}[\mathbf{E}[X_d|\mathbf{T}]] \geq ((1 - o(1))/\sqrt{2\pi d}) \sum_{v \in S_r} \mathbf{E}[\exp(-dg_v(r)^2/2)]$. Lemma 4.13 tells us that $\mathbf{E}[\exp(-dg_v(r)^2/2)] \geq v_4(c)/\sqrt{d}$. Hence, $\mathbf{E}[X_d] \geq \frac{v(c)}{d}\binom{\ell}{\ell/2 + \Gamma_d}$. □

4.6. Subtleties in Adding Noise

One might ask why we add noise in this particular fashion, and whether other ways of adding noise are equally effective. Since we only need ℓ random numbers, it seems intuitive that adding "more noise" could only help. For example, we might add noise on a per edge basis, that is, at each level i of *every* edge insertion, we choose a new random perturbation T_i of T. Interestingly, this version of noise does not smooth out the degree distribution, as shown in Figure 6. In this figure, the red curve corresponds to the degree distribution of the graph generated by NSKG with Graph500 parameters, $\ell = 26$, and $b = 0.1$. The blue curve corresponds to generation by adding noise per edge. As seen in this figure, adding noise per edge has hardly any effect on the oscillations, while NSKG provides a smooth degree distribution curve. (These results are fairly consistent over different parameter choices.) It is crucial that we use the *same* noisy $T_1, \ldots, T_\ell$ for every edge insertion.

5. EXPECTED NUMBER OF ISOLATED VERTICES

In this section, we give a simple formula for the number of isolated vertices in SKG. We focus on the symmetric case, where $t_2 = t_3$ in the matrix T. We assume that ℓ is even in the following, but the formula can be extended for ℓ being odd. The real contribution here is a clearer understanding of how many vertices SKG leaves isolated and how the SKG parameters affects this number.

THEOREM 5.1. *Consider SKG with T symmetric and let I denote the number of isolated vertices. With probability $1 - o(1)$,*

$$I = (1 \pm o(1)) \sum_{r=-\ell/2}^{r=\ell/2} \binom{\ell}{\ell/2 + r} \exp(-2\lambda\tau^r). \tag{3}$$

CLAIM 5.2. *Let q_r be the probability that a single edge insertion produces an in-edge or out-edge incident to $v \in S_r$. Then, for SKG with T symmetric,*

$$q_r = (1 \pm o(1)) \frac{2(1 - 4\sigma^2)^{\ell/2} \tau^r}{n}.$$

PROOF. Let $\mathcal{E}_o$ (resp. $\mathcal{E}_i$) be the event that a single edge insertion is an in-edge (resp. out-edge) of v. We have $q_r = \Pr(\mathcal{E}_o) + \Pr(\mathcal{E}_i) - \Pr(\mathcal{E}_o \cup \mathcal{E}_i)$. By Claim 3.3 and the symmetry to T, the first two probabilities are $\frac{(1 - 4\sigma^2)^{\ell/2} \tau^r}{n}$. The last is the probability that the edge insertion leads to a self-loop at v. This is at most $\sigma^\ell \Pr(\mathcal{E}_o)$. Since $\sigma < 1$, this is $o(\Pr(\mathcal{E}_o))$. □

As before, we can assume that $q_r \leq 1/\sqrt{m}$. By Claim 3.4, if $q_r \geq p_r \geq 1/\sqrt{m}$, then with probability tending to 1, vertices in slice r are not isolated. Hence, we can ignore such vertices when computing estimates for I.

CLAIM 5.3. *Let $v \in S_r$ and assume $q_r \leq 1/\sqrt{m}$. Then, for SKG with T symmetric,*

$$\Pr[v \text{ is isolated}] = (1 \pm o(1)) \exp(-2\lambda \tau^r).$$

PROOF. Using Claim 5.2 and $(1 - x)^m = (1 \pm o(1))e^{-xm}$, for $|x| \leq 1/\sqrt{m}$,

$$(1 - q_r)^m = (1 \pm o(1)) \exp(-2(1 \pm o(1))\Delta(1 - 4\sigma^2)^{\ell/2}\tau^r) = (1 \pm o(1)) \exp(-2(1 \pm o(1))\lambda \tau^r).$$

For large ℓ, this converges to $\exp(-2\lambda \tau^r)$. □

PROOF OF THEOREM 5.1. By Claim 5.3 and linearity of expectation, the expected number of isolated vertices is

$$(1 \pm o(1)) \sum_{r=-\ell/2}^{r=\ell/2} \binom{\ell}{\ell/2 + r} \exp(-2\lambda \tau^r).$$

To bound that actual number of isolated vertices, we use concentration inequalities for functions of independent random variables. Let Y denote the number of isolated vertices, and $X_1, X_2, \ldots, X_m$ be the labels of the m edge insertions. Note that all the X_i's are independent, and Y is some fixed function of $X_1, X_2, \ldots, X_m$. Suppose we fix all the edge insertions and just modify one insertion. Then, the number of isolated vertices can change by at most $c = 2$. Hence, the function defining Y satisfies a Lipschitz condition. This means that changing a single argument of Y (some X_i) modifies the value of Y by at most a constant (c). By McDiarmid's inequality [McDiarmid 1989],

$$\Pr[|Y - \mathbf{E}[Y]| > \epsilon] < 2 \exp\left(-\frac{2\epsilon^2}{c^2 m}\right).$$

Setting $\epsilon = \sqrt{m \log m}$, we get the probability that Y deviates from its expectation by more than $\sqrt{m \log m}$ is $o(1)$. The expected number of vertices is at least $\binom{\ell}{\ell/2} \exp(-2\lambda)$, and $\sqrt{m \log m}$ is a lower-order term with respect to this quantity. This completes the proof. □

The fraction of isolated vertices in a slice r is essentially $\exp(-\lambda \tau^r)$. Note that τ is larger than 1. Hence, this is a decreasing function of r. This is quite natural, since if a vertex v has many zeros in its representation (higher slice), then it is likely to have a larger degree (and less likely to be isolated). This function is doubly exponential in r, and therefore decreases quickly with r. The fraction of isolates rapidly goes to 0 (respectively, 1) as r is positive (respectively, negative).

Table V.
Percentage of isolated vertices with different noise
levels for the GRAPH500 parameters and $\ell = 26$.

Max. noise level (b)	% isolated vertices
0	51.12
0.05	49.26
0.06	49.12
0.07	49.06
0.08	49.07
0.09	49.16
0.1	49.34

5.1. Effect of Noise on Isolated Vertices

The introduction of noise was quite successful in correcting the degree distribution but has little effect on the number of isolated vertices. This is not surprising, considering the noise affects fat tail behavior of the degree distribution. The number of isolated vertices is a different aspect of the degree distribution. The data presented in Table V clearly shows that the number of isolated vertices is quite resistant to noise. While there is some decrease in the number of isolated vertices, this quantity is very small compared to the total number of isolated vertices. We have observed similar results on the other parameter settings.

In addition to this empirical study, we can also give some mathematical intuition behind these observations. The equivalent statement of Claim 5.3 for NSKG is

$$\Pr[v \text{ is isolated}] \geq (1 - o(1)) \exp(-2\lambda\tau^r) = (1 - o(1))[\exp(-2\lambda\tau^r)]^{\rho_v}.$$

The noiseless version of this probability is $[\exp(-2\lambda\tau_r)]$. Note that the probability now is a random variable that depends on T, since ρ_v depends on the noise. Lemma 4.13 tells us that $\ln \rho_v$ lies mostly in the range $[1 - c'/\sqrt{\ell}, 1 + c'/\sqrt{\ell}]$ (for constant c'), and is concentrated close to 1.

We are mainly interested in the case when the probability that v is isolated is *not* vanishingly small (is at least, say 0.01). As ℓ grows, ρ_v is close to being 1, and deviations are quite small. So, when we take the noiseless probability to the ρ_vth power, we get almost the same value.

5.2. Relation of SKG Parameters to the Number of Isolated Vertices

When λ decreases, the number of isolated vertices increases. Suppose we fix the SKG matrix and average degree Δ, and start increasing ℓ. Note that this is done in the Graph500 benchmark, to construct larger and larger graphs. The value of λ decreases exponentially in ℓ, so the number of isolated vertices will increase. Our formula suggests ways of counteracting this problem. The value of Δ could be increased, or the value σ could be decreased. But, in general, this will be a problem for generating large sparse graphs using a fixed SKG matrix.

When σ increases, then λ decreases and τ increases. Nonetheless, the effect of λ is much stronger than that of τ. Hence, the number of isolated vertices will increase as σ increases. In Table II, we compute the estimated number of isolated vertices in graphs for the Graph500 parameters. Observe how the fraction of isolated vertices consistently increases as ℓ is increased. For the largest setting of $k = 42$, only one fourth of the vertices are not isolated.

6. *K*-CORES IN SKG

Structures of k-cores are an important part of social network analysis [Carmi et al. 2007; Alvarez-Hamelin et al. 2008; Kumar et al. 2010], as they are a manifestation of the community structure and high connectivity of these graphs.

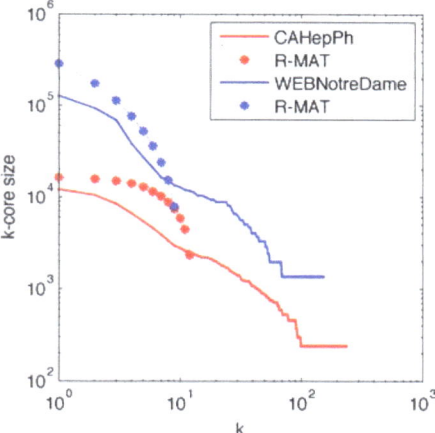

Fig. 7. Core decompositions of real graphs and their SKG model. Observe that the max core of SKG is an order of magnitude smaller.

Definition 6.1. Given an undirected graph $G = (V, E)$, the *subgraph induced by set* $S \subseteq V$, is denoted by $G|_S := (S, E')$, where E' contains every edge of E that is completely contained in S. For an undirected graph, the *k-core* of G the largest induced subgraph of minimum degree k. The *max core number* of G is the largest k such that G contains a (non-empty) k-core. (These can be extended to directed versions: a k-out-core is a subgraph with min out-degree k.)

A bipartite core is an induced subgraph with every vertex has *either* a high in-degree or out-degree. The former are called *authorities* and the latter are *hubs*. Large bipartite cores are present in web graphs and are an important structural component [Gibson et al. 1998; Kleinberg 1999]. Note that if we make the directed graph undirected (by simply removing the directions), then a bipartite core becomes a normal core. Hence, it is useful to compute cores in a directed graph by making it undirected.

We begin by comparing the sizes of k-cores in real graphs, and their models using SKG [Leskovec et al. 2010]. Refer to Figure 7. We plot the size of the maximum k-core with k. The k at which the curve ends is the max core number. (For CAHepPh, we look at undirected cores, since this is an undirected graph. For WEBNotreDame, a directed graph, we look at out-cores. But the empirical observations we make holds for all other core versions.) For both our examples, we see how drastically different the curves are. By far the most important difference is that the curve for the SKG versions are extremely short. This means that the max core number is *much smaller* for SKG modeled graphs compared to their real counterparts. For the web graph WEBNotreDame, we see the presence of large cores, probably an indication of some community structure. The maximum core number of the SKG version is an *order of magnitude* smaller. Minor modifications (like increasing degree, or slight variation of parameters) to these graphs do not increase the core sizes or max cores numbers much. This is a problem, since this is strongly suggesting that SKG do not exhibit localized density like real web graphs or social networks.

If we wish to use SKG to model real networks, then it is imperative to understand the behavior of max core numbers for SKG. Indeed, in Table VI, we see that our observation is not just an artifact of our examples. SKG consistently have very low max core number. Only for the peer-to-peer Gnutella graphs does SKG match the real data, and this is specifically for the case where the max core number is extremely small. For

Table VI. Core Sizes in Real Graphs and SKG Version

Graph	Real max core	SKG max core
CAGrQc	43	4
CAHepPh	238	16
CAHepTh	31	5
CITHepPh	30	19
CITHepTh	37	19
P2PGnutella25	5	5
P2PGnutella30	7	6
SOCEpinions	67	43
WEBNotreDame	155	31

the undirected graph (the first three co-authorship networks), we have computed the undirected cores. The corresponding SKG is generated by copying the upper triangular part in the lower half to get a symmetric matrix (an undirected graph). The remaining graphs are directed, and we simply remove the direction on the edges and compute the total core. Our observations hold for in and out cores as well, and for a wide range of data. This is an indication that SKG is not generating sufficiently dense subgraphs.

We focus our attention on the max core number of SKG. How does this number change with the various parameters? The following summarizes our observations.

Empirical Observation 6.2. For SKG with symmetric T, we have the following observations:

(1) The max core number increases with σ. By and large, if $\sigma < 0.1$, max core numbers are extremely tiny.
(2) Max core numbers grow with ℓ only when the values of σ are sufficiently large. Even then, the growth is much slower than the size of the graph. For smaller σ, max core numbers exhibit essentially negligible growth.
(3) Max core numbers increase essentially linearly with Δ.

Large max core numbers require larger values of σ. As mentioned in Section 5, increasing σ increases the number of isolated vertices. Hence, there is an inherent tension between increasing the max core number and decreasing the number of isolated vertices.

For the sake of consistency, we performed the following experiments on the max core after taking a symmetric version of the SKG graph. Our results look the same for in and out cores as well. In Figure 8(a), we show how increasing σ increases the max core number. We fix the values of $\ell = 16$ and $m = 6 \times 2^{16}$. (There is nothing special about these values. Indeed the results are basically identical, regardless of this choice.) Then, we fix t_1 (or t_2) to some value, and slowly increase σ by increasing t_2 (resp. t_1). We see that regardless of the fixed values of t_1 (or t_2), the max core consistently increases. But as long as $\sigma < 0.1$, max core numbers remain almost the same.

In Figure 8(b), we fix matrix T and average degree Δ, and only vary ℓ. For WEB-NotreDame,[5] we have $\sigma = 0.18$ and for CA-HEP-Ph, we have $\sigma = 0.11$. For both cases, increasing ℓ barely increases the max core number. Despite increasing the graph size by 8 orders of magnitude, the max core number only doubles. Contrast this with the Graph500 setting, where $\sigma = 0.26$, and we see a steady increase with larger ℓ. This is a predictable pattern we notice for many different parameter settings: larger σ leads to larger max core numbers as ℓ goes up. Finally, in Figure 8(c), we see that the max core number is basically linear in Δ.

---

[5]Even though the matrix T is not symmetric, we can still define σ. Also, the off diagonal values are 0.20 and 0.21, so they are almost equal.

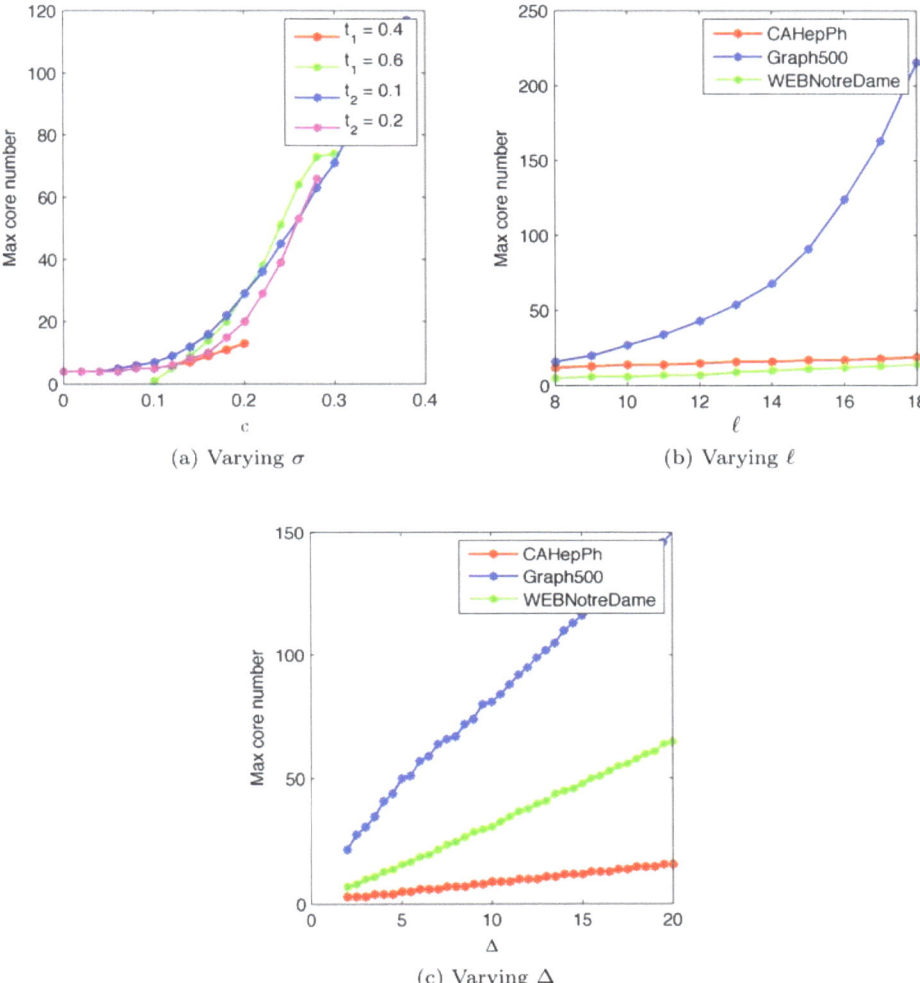

Fig. 8. We plot the max core number against various parameters. In the first picture, we plot the max core number of an (symmetric) SKG graph with increasing σ. Next, we show how the max core number increases with ℓ, the number of levels. Observe the major role that the matrix σ plays. For Graph500, σ is much larger than the other parameter sets. Finally, we show that regardless of the parameters, the max core number increases linearly with Δ.

6.1. Effect of Noise on Cores

Our general intuition is that NSKG mainly redistributes edges of SKG to get a smooth degree distribution, but does not have major effects on the overall structure of the graph. This is somewhat validated by our studies on isolated vertices and reinforced by looking at k-cores. In Figure 9, we plot the core decompositions of SKG and two versions on NSKG ($b = 0.05$ and $b = 0.1$). We observe that there are little changes in these decompositions, although there is a smoothening of the curve for Graph500 parameters. The problem of tiny cores of SKG is not mitigated by the addition of noise.

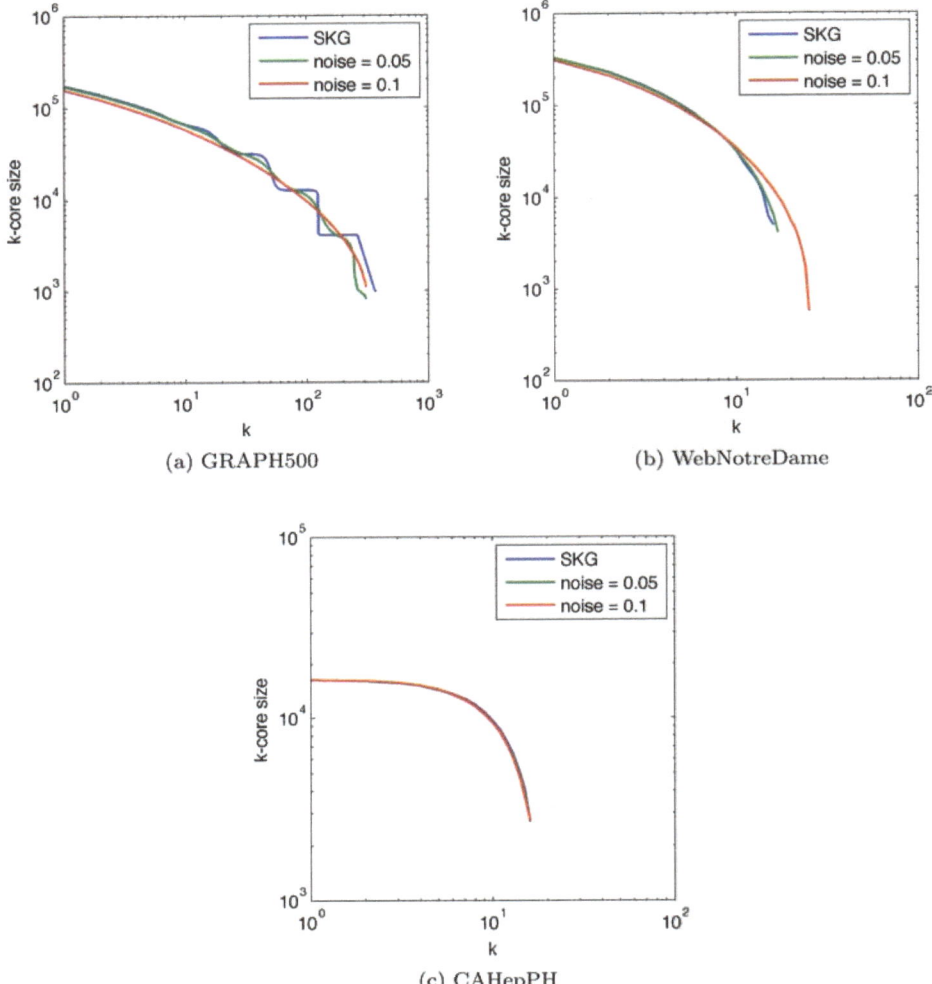

Fig. 9. We plot the core decomposition of SKG and NSKG (with 2 settings of noise) for the different parameters. Observe that there is only a minor change in core sizes with noise.

7. CONCLUSIONS

For a true understanding of a model, a careful theoretical and empirical study of its properties in relation to its parameters is imperative. This not only provides insight into why certain properties arise, but also suggests ways for enhancement. One strength of the SKG model is its amenability to rigorous analysis, which we exploit in this article.

We prove strong theorems about the degree distribution, and more significantly show how adding noise can give a true lognormal distribution by eliminating the oscillations in degree distributions. Our proposed method of adding noise requires only ℓ random numbers all together, and is hence cost effective. We want to stress that our major contribution is in providing *both* the theory and matching empirical evidence. The formula for expected number of isolated vertices provides an efficient alternative to methods for computing the full degree distribution. Besides requiring fewer operations to compute and being less prone to numerical errors, the formula transparently relates the expected number of isolated vertices to the SKG parameters. Our studies on core numbers establish a connection between the model parameters and the cores of the resulting graphs. In particular, we show that commonly used SKG parameters generate tiny cores, and the model's ability to generate large cores is limited.

ACKNOWLEDGMENT

We are grateful to David Gleich for the MATLAB BGL library as well as many helpful discussions. We thank Todd Plantenga for creating large SKG and NSKG instances, and for generating Figure 1. We also thank Jon Berry for checking our Graph500 predictions against real data, and also David Bader and Richard Murphy for discussions about the Graph500 benchmark. We acknowledge the inspiration of Jennifer Neville and Blair Sullivan, who inspired us with their different work on SKG during recent visits to Sandia.

REFERENCES

ALVAREZ-HAMELIN, J. I., DALL'ASTA, L., BARRAT, A., AND VESPIGNANI, A. 2008. K-core decomposition of internet graphs: hierarchies, self-similarity and measurement biases. *Netw. Hetero. Media 3*, 2, 371–393.

ANDERSEN, R. AND CHELLAPILLA, K. 2009. Finding dense subgraphs with size bounds. In *Algorithms and Models for the Web-Graph*, Springer, 25–37.

BERRY, A. 1941. The accuracy of the gaussian approximation to the sum of independent variates. *Trans. AMS 49*, 1, 122–136.

BI, Z., FALOUTSOS, C., AND KORN, F. 2001. The "DGX" distribution for mining massive, skewed data. In *Proceedings of KDD '01*. ACM, 17–26.

CARMI, S., HAVLIN, S., KIRKPATRICK, S., SHAVITT, Y., AND SHIR, E. 2007. A model of internet topology using k-shell decomposition. *Proc. Nat. Acad. Sci. 104*, 27, 11150–11154.

CHAKRABARTI, D. AND FALOUTSOS, C. 2006. Graph mining: Laws, generators, and algorithms. *ACM Comput. Surv. 38*, 1.

CHAKRABARTI, D., ZHAN, Y., AND FALOUTSOS, C. 2004. R-MAT: A recursive model for graph mining. In *Proceedings of SDM '04*. 442–446.

CLAUSET, A., SHALIZI, C. R., AND NEWMAN, M. E. J. 2009. Power-law distributions in empirical data. *SIAM Rev. 51*, 4, 661–703.

ESSEEN, C.-G. 1942. A moment inequality with an application to the central limit theorem. *Skand. Aktuari-etidskr 39*, 160–170.

FELLER, W. 1968. *An Introduction to Probability Theory and Applications: Vol I.* 3rd Ed. Wiley.

GIBSON, D., KLEINBERG, J., AND RAGHAVAN, P. 1998. Inferring web communities from link topology. In *Proceedings of HYPERTEXT '98*. ACM, 225–234.

GKANTSIDIS, C., MIHAIL, M., AND ZEGURA, E. W. 2003. Spectral analysis of internet topologies. In *Proceedings of INFOCOM 2003*. IEEE, 364–374.

GOLTSEV, A. V., DOROGOVTSEV, S. N., AND MENDES, J. F. F. 2006. *k*-core (bootstrap) percolation on complex networks: Critical phenomena and nonlocal effects. *Phys. Rev. E 73*, 5, 056101.

GRAPH500 STEERING COMMITTEE. 2010. Graph 500 benchmark. http://www.graph500.org/specifications.

GROËR, C., SULLIVAN, B. D., AND POOLE, S. 2011. A mathematical analysis of the R-MAT random graph generator. *Networks 58*, 3, 159–170.

IBRAGIMOV, I. A. 1956. On the composition of unimodal distributions. *Theory Probability its App. 1*, 2, 255–260.

KIM, M. AND LESKOVEC, J. 2010. Multiplicative attribute graph model of real-world networks. arXiv:1009.3499v2.

KLEINBERG, J. M. 1999. Authoritative sources in a hyperlinked environment. *J. ACM 46*, 5, 604–632.

KUMAR, R., NOVAK, J., AND TOMKINS, A. 2010. Structure and evolution of online social networks. In *Link Mining: Models, Algorithms, and Applications*, Springer, 337–357.

LESKOVEC, J., CHAKRABARTI, D., KLEINBERG, J., AND FALOUTSOS, C. 2005. Realistic, mathematically tractable graph generation and evolution, using Kronecker multiplication. In *Proceedings of PKDD 2005*. Springer, 133–145.

LESKOVEC, J., CHAKRABARTI, D., KLEINBERG, J., FALOUTSOS, C., AND GHAHRAMANI, Z. 2010. Kronecker graphs: An approach to modeling networks. *J. Mach. Learn. Res. 11*, 985–1042.

LESKOVEC, J. AND FALOUTSOS, C. 2007. Scalable modeling of real graphs using kronecker multiplication. In *Proceedings of ICML '07*. ACM, 497–504.

MAHDIAN, M. AND XU, Y. 2007. Stochastic kronecker graphs. In *Algorithms and Models for the Web-Graph*, Springer, 179–186.

MAHDIAN, M. AND XU, Y. 2011. Stochastic Kronecker graphs. *Rand. Struct. Algor. 38*, 4, 453–466.

McDIARMID, C. 1989. On the method of bounded differences. *Surv. Combinat. 141*, 148–188.

MILLER, B., BLISS, N., AND WOLFE, P. 2010. Subgraph detection using eigenvector L1 norms. In *Proceedings of NIPS 2010*. 1633–1641.

MITZENMACHER, M. 2003. A brief history of generative models for power law and lognormal distributions. *Internet Math. 1*, 2, 226–251.

MITZENMACHER, M. 2006. The future of power law research. *Internet Math. 2*, 4, 525–534.

MORENO, S., KIRSHNER, S., NEVILLE, J., AND VISHWANATHAN, S. V. N. 2010. Tied Kronecker product graph models to capture variance in network populations. In *Proceedings of the 48th Annual Allerton Conference on Communication, Control, and Computing*. 1137–1144.

MOTWANI, R. AND RAGHAVAN, P. 1995. *Randomized Algorithms*. Cambridge University Press.

PENNOCK, D., FLAKE, G., LAWRENCE, S., GLOVER, E., AND GILES, C. L. 2002. Winners don't take all: Characterizing the competition for links on the web. *Proc. Nat. Acad. Sci. 99*, 8, 5207–5211.

SALA, A., CAO, L., WILSON, C., ZABLIT, R., ZHENG, H., AND ZHAO, B. Y. 2010. Measurement-calibrated graph models for social network experiments. In *Proceedings of WWW '10*. ACM, 861–870.

Received September 2011; revised August 2012 and December 2012; accepted December 2012

www.ingramcontent.com/pod-product-compliance
Lightning Source LLC
Chambersburg PA
CBHW050411180526
45159CB00005B/2229